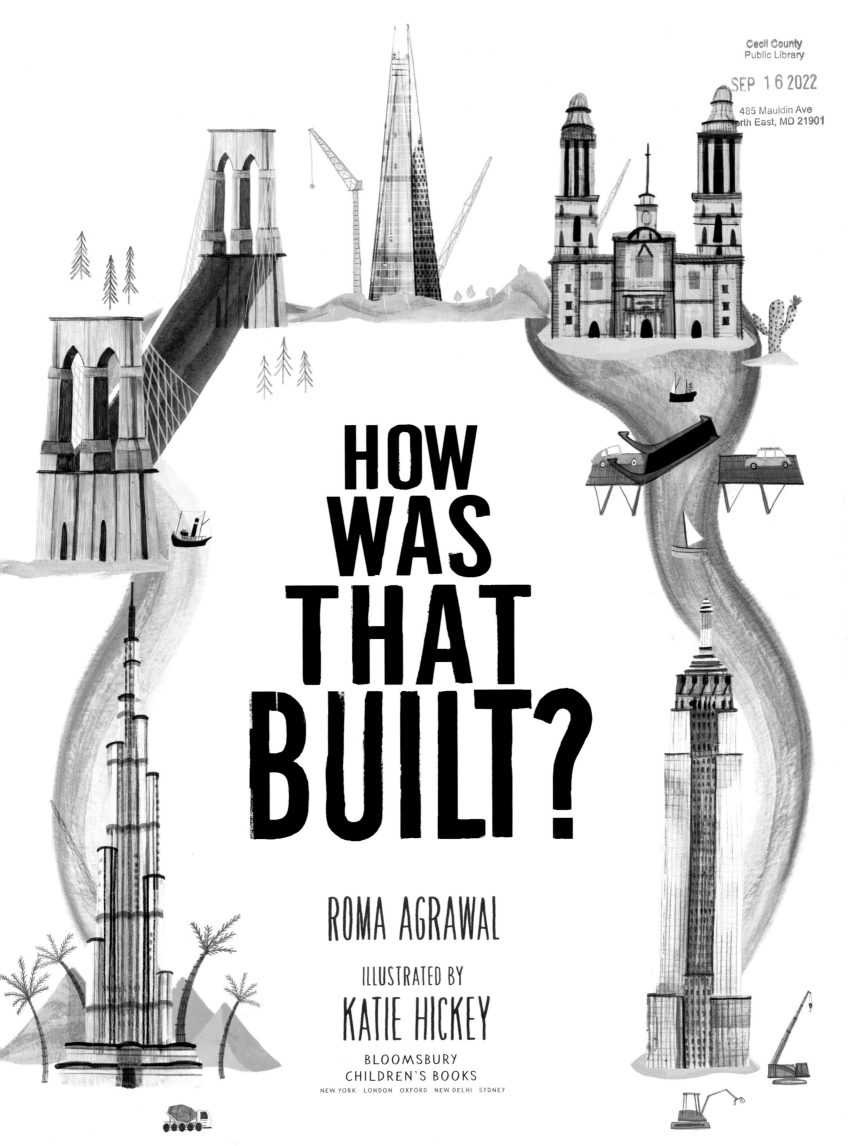

HOW WAS THAT BUILT?

ROMA AGRAWAL

ILLUSTRATED BY

KATIE HICKEY

BLOOMSBURY
CHILDREN'S BOOKS

NEW YORK   LONDON   OXFORD   NEW DELHI   SYDNEY

For Zarya and Kiah —R. A.

For my Dad, forever tinkering and forever building —K. H.

BLOOMSBURY CHILDREN'S BOOKS
Bloomsbury Publishing Inc., part of Bloomsbury Publishing Plc
1385 Broadway, New York, NY 10018

BLOOMSBURY, BLOOMSBURY CHILDREN'S BOOKS, and the Diana logo
are trademarks of Bloomsbury Publishing Plc

First published in Great Britain in September 2021 by Bloomsbury Publishing Plc
Published in the United States of America in August 2022
by Bloomsbury Children's Books

Bloomsbury books may be purchased for business or promotional use. For information on bulk
purchases please contact Macmillan Corporate and Premium Sales Department at
specialmarkets@macmillan.com

Library of Congress Cataloging-in-Publication Data
Names: Agrawal, Roma, 1983- author. | Hickey, Katie, illustrator.
Title: How was that built? : the stories behind awesome structures / Roma Agrawal, Katie Hickey.
Description: New York : Bloomsbury Children's Books, 2022. | Audience: Grades 2–3
Summary: From skyscrapers to bridges, meet the extraordinary people who helped
build some of the world's architectural marvels—Provided by publisher.
Identifiers: LCCN 2021045085 |
ISBN 978-1-5476-0929-1 (hardcover)
Subjects: LCSH: Structural engineering—Juvenile works. | Building—Juvenile works.
Classification: LCC TA634 .A45 2022 | DDC 624.1—dc23
LC record available at https://lccn.loc.gov/2021045085

Book design by Katie Knutton
Typeset in Korolev Compressed, Amorie SC, and Museo Sans
Printed in China by Leo Paper Products, Heshan, Guangdong
2 4 6 8 10 9 7 5 3 1

To find out more about our authors and books visit www.bloomsbury.com and sign up for our newsletters.

CONTENTS

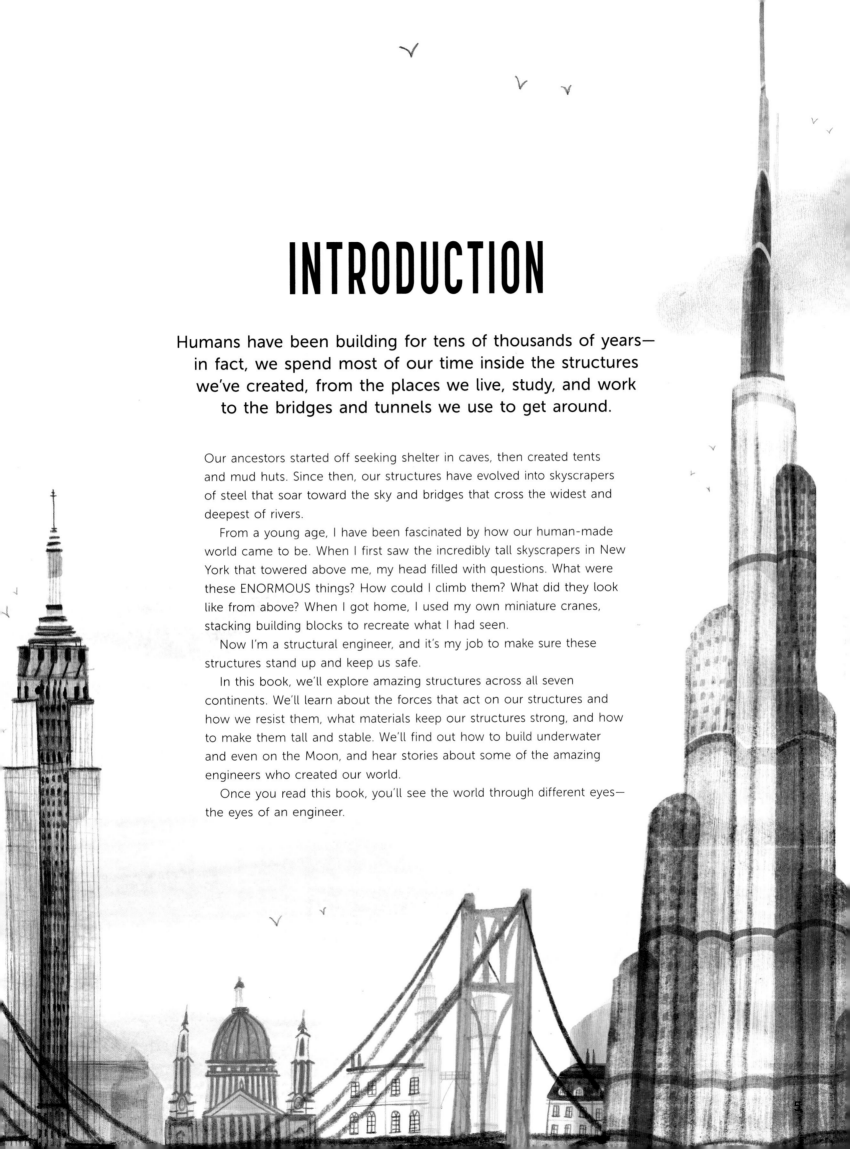

INTRODUCTION

Humans have been building for tens of thousands of years—
in fact, we spend most of our time inside the structures
we've created, from the places we live, study, and work
to the bridges and tunnels we use to get around.

Our ancestors started off seeking shelter in caves, then created tents
and mud huts. Since then, our structures have evolved into skyscrapers
of steel that soar toward the sky and bridges that cross the widest and
deepest of rivers.

From a young age, I have been fascinated by how our human-made
world came to be. When I first saw the incredibly tall skyscrapers in New
York that towered above me, my head filled with questions. What were
these ENORMOUS things? How could I climb them? What did they look
like from above? When I got home, I used my own miniature cranes,
stacking building blocks to recreate what I had seen.

Now I'm a structural engineer, and it's my job to make sure these
structures stand up and keep us safe.

In this book, we'll explore amazing structures across all seven
continents. We'll learn about the forces that act on our structures and
how we resist them, what materials keep our structures strong, and how
to make them tall and stable. We'll find out how to build underwater
and even on the Moon, and hear stories about some of the amazing
engineers who created our world.

Once you read this book, you'll see the world through different eyes—
the eyes of an engineer.

HOW TO BUILD FLAT
THE METROPOLITAN CATHEDRAL

Before we build structures, we need to study the ground we're building on. If we don't have strong foundations, buildings can sink or tilt. The Leaning Tower of Pisa in Italy is a famous example of things going wrong because the ground was soft and the structure moved in unexpected ways. So imagine what happens when the city you're building is in a lake.

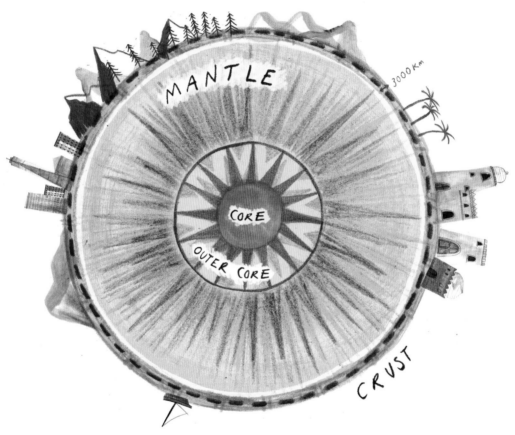

Digging deep

The Earth is made up of different layers. The top layer, the crust, averages about 18 miles thick. This is what our continents and the bottom of our oceans are made from. The layer below is called the mantle, which is much thicker—nearly 2,000 miles thick. Next comes the outer core, which is so hot that it has melted into a liquid, and finally, in the center, is the solid inner core. If you dug a hole from the surface to the center of the Earth, it would almost cover the same distance as flying from Chicago, IL, to London, UK.

Different types of soils

Our buildings have foundations that are built in the Earth's crust. Some of the deepest foundations go about 160 yards down into the ground, which is only a tiny portion of this layer.

The crust below us is made of different types of soils all around the world. In places like deserts, there is a thick layer of sand on top of rock. Near rivers, the soil is wet and soft because the water seeps in. Some places have gravel, which is a mix of different-sized stones, on top of clay, on top of sand. This makes the construction of every structure different, as each type of soil needs a specific foundation and approach.

Before we design foundations, we need to know what type of soil we're building on. One way to find out is to use special maps that tell us the history of the ground. We can dig shallow holes to check the soil just below the surface. And to look really far down, we use a borehole, or deep well, where we dig out about a 100-foot-deep section of earth and examine the layers in the ground it reveals.

FOUNDATIONS AND FORCES

Rafts

Rafts are used to build on softer soils. These are thick slabs of concrete or other strong material that float on top of the ground. Like a boat—which spreads our weight across water and stops us from sinking—a raft distributes the weight of a structure across the ground.

But if a structure is too heavy or the soil is too soft, rafts can sink. Then, we use piles. These are long columns we put in the ground. One way of constructing piles is to use a piling rig, a machine with a giant corkscrew that can scoop out a long thin hole in the ground. Then this hole is filled with concrete. Piles used to be made from tree trunks. These days, as an alternative to concrete, steel or timber may be used.

Piles

Piles are clever because they can work in two ways, using friction and bearing. If you try sliding your foot on a wooden floor when you're wearing socks, it's really easy to move around. But if you wear shoes, your foot sticks to the floor and it's tricky to slide. That's because shoes increase the stickiness, or friction, between your foot and the floor. Piles create a friction force between themselves and the soil, which stops the building from sinking.

The trick is to make sure there is enough friction!

If the friction force isn't strong enough to hold the building up, then we can dig our piles deep down until we find a very strong layer of sand or rock to support it. Here, the piles push down on this material directly without sinking any deeper. This is called a bearing pile.

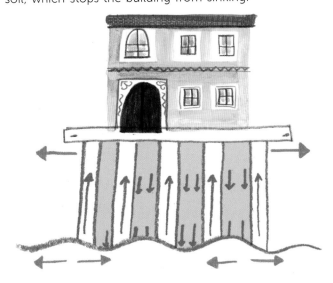

A city on a lake

In 1325, the Aztecs, a people in what is today Mexico, founded their new city, Tenochtitlan, in the middle of Lake Texcoco. Why there? They had a vision: their god Huitzilopochtli (god of war and sun) had said their new capital city must be built where they found an eagle with a snake in its beak sitting on top of a nopal cactus. The Aztecs roamed for over 250 years, and when they finally saw the sign, the cactus with the eagle and snake was on a small island in the middle of a lake! They built a beautiful city in part of the lake by filling it in with earth and building platforms on timber piles. It was a scenic place with canals and large pyramid temples. Its rulers commanded vast lands.

Mexico City

Spanish invaders captured Tenochtitlan in 1521. They destroyed it and started to build a new city there. They filled in the rest of the lake and built many structures, including a cathedral.

Mexico City in Mexico, built over the ruins of Tenochtitlan, has grown quite a lot since that time, but its center is still on top of the filled-in lake. The soil here is very wet and weak, so the buildings are sinking, fast. It is like a bowl full of jelly with buildings on top. Over the past 150 years, this area has sunk by more than the height of a three-story building!

A sinking cathedral

The Metropolitan Cathedral, at the center of Mexico City, is one building that sank and tilted because of the soft ground. It was leaning and parts of it were in danger of collapse. In the 1990s, engineers had to save this structure.

The cathedral sits on top of an old Aztec pyramid and also the filled-in lake. Spanish engineers built a raft foundation to spread the structure's weight over the soil, but the soil was too soft and it started to sink unevenly.

Imagine you have a bowl of sand with a coaster sitting on top. If you push down on one corner of the coaster, you'll see the sand below get squashed and the coaster sit crookedly. To make the coaster flat again, you could do one of two things: you could push down on the opposite corner of the coaster or you could remove some of the sand below the higher corner of the coaster until it straightens out.

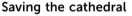

Saving the cathedral

If your coaster is a massive cathedral, you can't push down with enough weight to straighten the cathedral! So, ingenious engineers dug 32 huge shafts, like wells, below the foundation. Then they pumped water out, drilled long, thin holes spreading out horizontally from each shaft, and removed soil. Where the cathedral had tilted upwards the most, they removed the most soil. This way the cathedral tilted in the opposite direction and became more level than before.

What next?

Measurements are being taken all the time to check how much the building is sinking. The good news is that it is going down very slowly, and not tilting any more. The cathedral has been saved! Everything the engineers learned from saving the Metropolitan Cathedral can be used by future engineers, especially for building in harsh conditions as populations expand and the climate changes.

A giant, missile-shaped pendulum directly under the central dome shows how far the cathedral has shifted.

Sensors in glass boxes are positioned throughout the cathedral. These send data wirelessly to a lab in Italy where engineers monitor how the structure is behaving.

Temporary steel beams and props supported the cathedral's arches and columns to prevent damage from sudden movements to the structure while the engineers worked on the soil.

32 large, long holes were dug by hand underneath the cathedral, through the foundation and into the ground, so that engineers could access the soil below the cathedral.

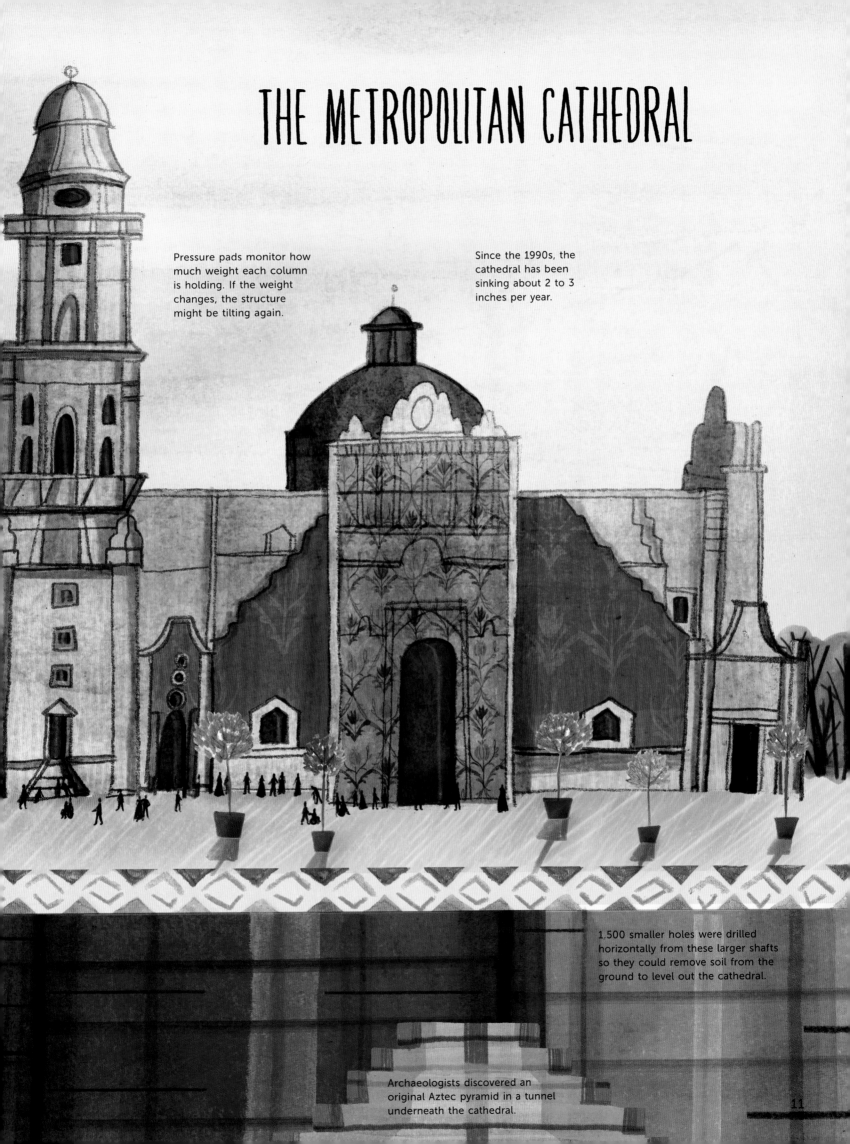

THE METROPOLITAN CATHEDRAL

Pressure pads monitor how much weight each column is holding. If the weight changes, the structure might be tilting again.

Since the 1990s, the cathedral has been sinking about 2 to 3 inches per year.

1,500 smaller holes were drilled horizontally from these larger shafts so they could remove soil from the ground to level out the cathedral.

Archaeologists discovered an original Aztec pyramid in a tunnel underneath the cathedral.

11

HOW TO
BUILD TALL
THE SHARD

The Shard is the tallest tower in western Europe. It has a distinct triangular shape and is near the River Thames in London, UK. Tall buildings are challenging and interesting to design. There are different forces in nature that we need to resist to make sure skyscrapers stay standing and don't collapse.

What makes a building stand?

Gravity attracts everything toward the Earth's center—that's why when we throw a ball up into the air, it falls back down. Gravity also pulls down on all our structures. It's the engineers' job to make sure the structure's framework is made from the right materials and is strong enough to fight this force.

The main framework for buildings consists of horizontal beams, which make up the floors, ceilings, and roofs, and vertical columns, which hold up the beams and form walls. In a skyscraper, we have to calculate how much the materials it's made from will weigh, and also how much the stuff inside it will weigh, from elevators and air-conditioning units, books, computers, and desks to all the people! We can then do the math to check that the steel or concrete beams and columns won't get crushed by this weight and that our skyscrapers will reach brilliant heights.

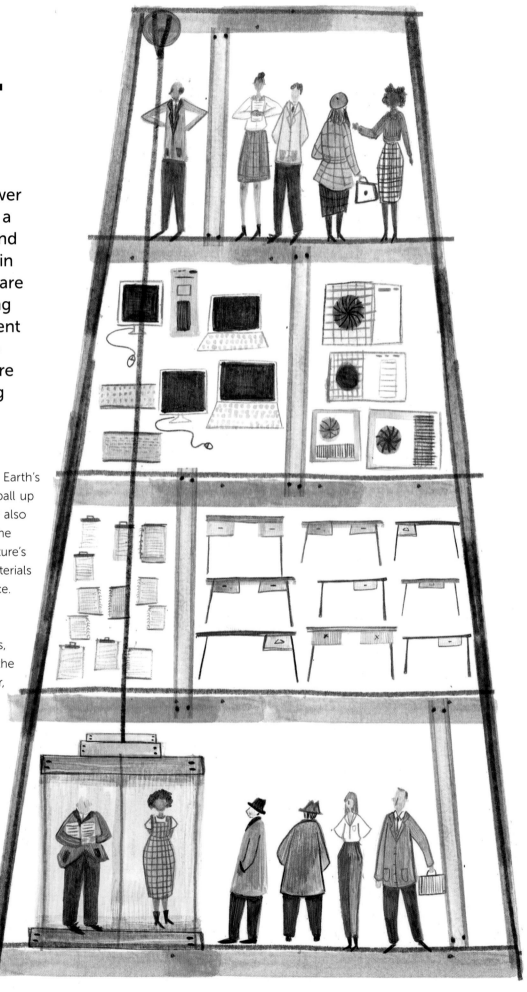

Beams

Imagine holding a carrot lengthwise between your hands and bending the ends up to form a U-shape. The top side gets squashed and the bottom gets pulled apart. Engineers call the squashing force compression and the pulling force tension. When the tension is large enough, the carrot snaps. When the compression is large enough, the top crushes. This is how beams work. Engineers check the forces acting on beams to make sure they won't move too much or break.

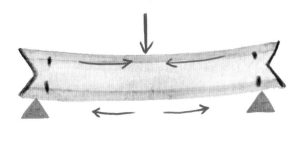

Columns

Try these two simple experiments to see how columns can fail. Roll up a piece of paper into a tube and tape it together. Stand it up on a table and put a small, light book on it. You'll see the tube is strong enough to hold the book up: that's what a good column does. But if you put a really heavy book on top, the tube will crush and the book will fall down. That's a bad column, which has failed by crushing. To hold up the heavy book, you would need a much stronger tube. Columns can also fail by bending. If you hold a ruler vertically on a table and push down on it, you will see it bowing. Don't push too hard or your ruler will snap!

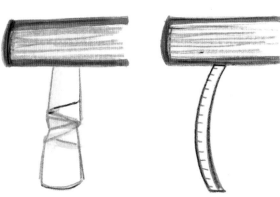

Steel for strength

Millions of tiny atoms, arranged in patterns to form crystals, make up metals such as iron and steel. The earliest metal used in big buildings was wrought iron. But this is a relatively soft metal because its crystals slide around a little when pushed and pulled. To make iron stronger, engineers added carbon. The atoms of carbon sat within the iron crystals and stopped them from moving as much, making a metal called steel. When you try to pull steel apart, the crystals don't move as easily, and so it's a stronger material for building. But you need the perfect amount of carbon: too much makes metals brittle, which means they can crack easily.

TRY IT AT HOME: STEEL

Take a large plate and pour some malted milk balls onto it. Roll your palm over them. You'll see that the chocolates move around easily—this is like the crystals of pure iron. Now sprinkle some raisins between the malted milk balls and try again. The raisins block the malted milk balls from rolling around as easily, which is how carbon atoms make steel stronger.

How do we make steel?

To build upward, first we have to gather materials deep inside the earth. Iron mined from the ground has a mix of different impurities, such as carbon, silicon, and phosphorus. A British engineer called Henry Bessemer invented a process for making steel cheaply in the 19th century. He put iron pieces into a covered furnace and blew hot air into it. A chemical reaction happened. The oxygen in the air reacted with the carbon in the iron and released huge amounts of heat. This heat took away the impurities and left behind pure iron. Then Bessemer could add in the exact quantity of carbon needed to make the best steel. Since then, steel has been used all over the world to build our most exciting buildings, bridges, stadiums, and railways.

Henry Bessemer

THE SHARD

Renzo Piano

Renzo Piano is a famous architect who designed The Shard. He also worked on the Pompidou Center in Paris. His design for The Shard was inspired by the spires of London churches and the masts of tall ships.

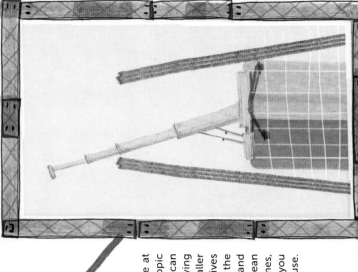

The top part of The Shard, called the Spire, is made from steel. The steel beams and columns in the Spire are carefully joined together with bolts (which are like large screws) and welds (this is where a hot flame melts some steel to stick different pieces together).

The Shard has 87 stories, with an observation deck on the 72nd floor.

On July 5, 2012, an amazing laser and spotlight display celebrated the completion of the outside of The Shard.

The open steel structure in the Spire has special paint to protect it from the wind and sun.

Five tower cranes were used during construction.

This final crane at level 87 is a telescopic crane, which means it can be expanded for carrying out jobs and made smaller for storing away. It lives permanently at the top of The Shard and is still used to clean windows. Sometimes, if you're lucky, you can see it in use.

The Shard is over 1,000 feet tall, and its core is made from concrete. The core is at the center of a tall building and makes sure that it can resist wind forces.

The Shard has 11,000 glass panels, which would cover eight football fields!

The Shard was built using cranes that were cleverly assembled at various points during construction. One of these was on top of the core of the tower. As the core grew taller, this crane was simply able to rise with it.

Another crane, which was attached about halfway up the tower, was used to continue building right up to the steel Spire at the very top. To take the crane down after it finished its work, a smaller crane was built at level 72— but how did we take this one down? We used yet another crane, this one at level 87!

Concrete was used in the hotel and apartment zone, where there are lots of walls and different rooms. Here, it's better to use concrete because the concrete floor is thinner and saves space, and the concrete also absorbs more sound so guests can sleep better!

Steel beams, which are strong in tension and can go a long way without columns supporting them, were used for office levels where big open spaces were needed without too much structure.

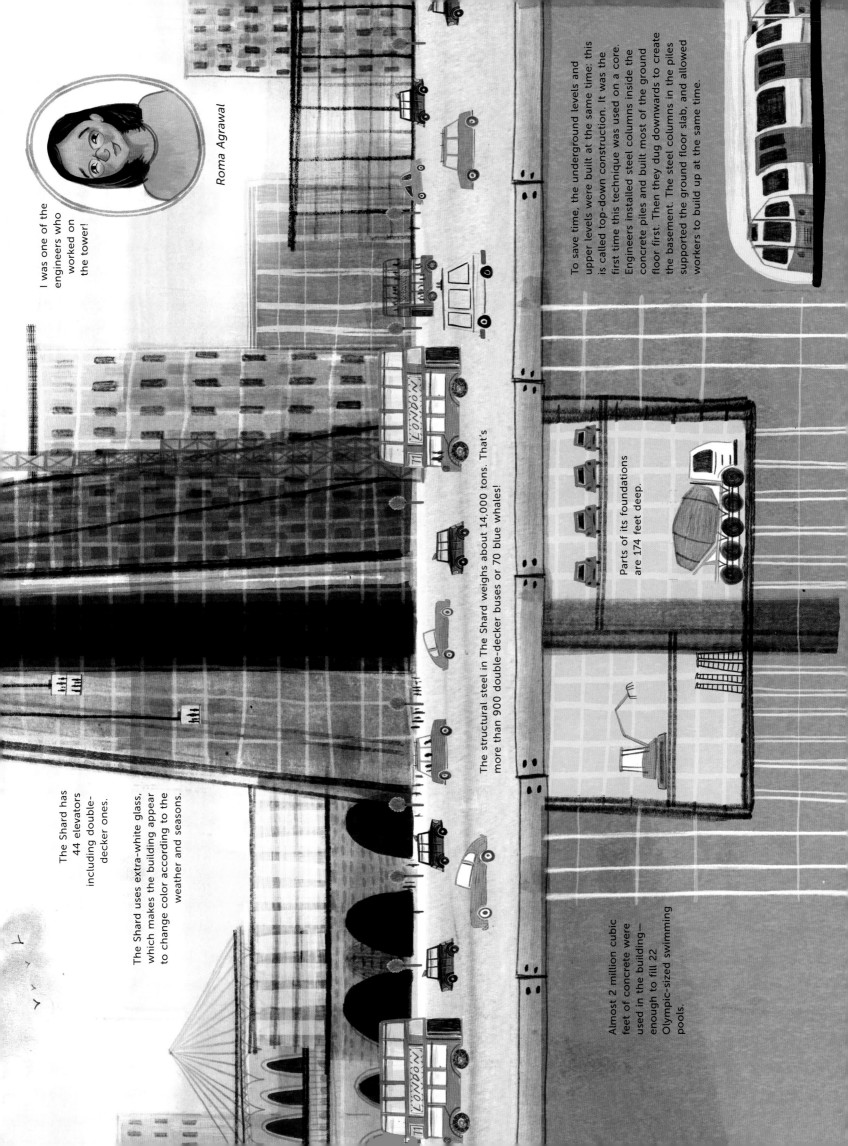

I was one of the engineers who worked on the tower!

Roma Agrawal

The Shard has 44 elevators including double-decker ones.

The Shard uses extra-white glass, which makes the building appear to change color according to the weather and seasons.

The structural steel in The Shard weighs about 14,000 tons. That's more than 900 double-decker buses or 70 blue whales!

Almost 2 million cubic feet of concrete were used in the building—enough to fill 22 Olympic-sized swimming pools.

Parts of its foundations are 174 feet deep.

To save time, the underground levels and upper levels were built at the same time: this is called top-down construction. It was the first time this technique was used on a core. Engineers installed steel columns inside the concrete piles and built most of the ground floor first. Then they dug downwards to create the basement. The steel columns in the piles supported the ground floor slab, and allowed workers to build up at the same time.

THE MACHINES THAT BUILD TALL BUILDINGS

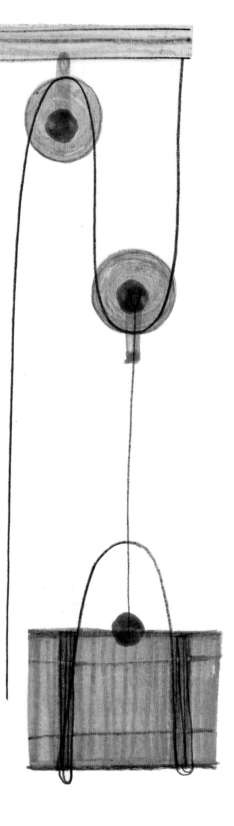

To build tall, we have to lift materials up high, and we use machines called cranes to do this. Modern cranes are made of crisscrossing pieces of steel with a large arm at the top. At the end of this arm, or jib, are strong steel ropes and a hook to which we attach the heavy items that need to be moved. The ropes are wound back through pulleys, which lift the hook up along with the materials.

What is a pulley?

A pulley is a wheel with a rope wrapped around it. One end of the rope is tied to a heavy object that needs to be lifted, while a person pulls, or applies force, on the other end. Ancient civilizations used pulleys to raise a bucket full of water from a well, and humans have found lots of uses for pulleys since then.

But to lift very heavy objects, we need the ancient Greek scientist Archimedes's invention called the compound pulley, which is made up of lots of pulleys with a single rope weaving its way through them. This helped lift heavier loads because the person pulling on the other end of the rope didn't need to use as much effort. If you have two pulleys instead of one, you only need half the amount of force. This made it much easier to lift materials high into the sky. The Romans used compound pulleys in their cranes, and were able to build tall structures—such as their multistory apartment buildings called *insulae*—thanks to this invention.

Archimedes

TYPES OF CRANES

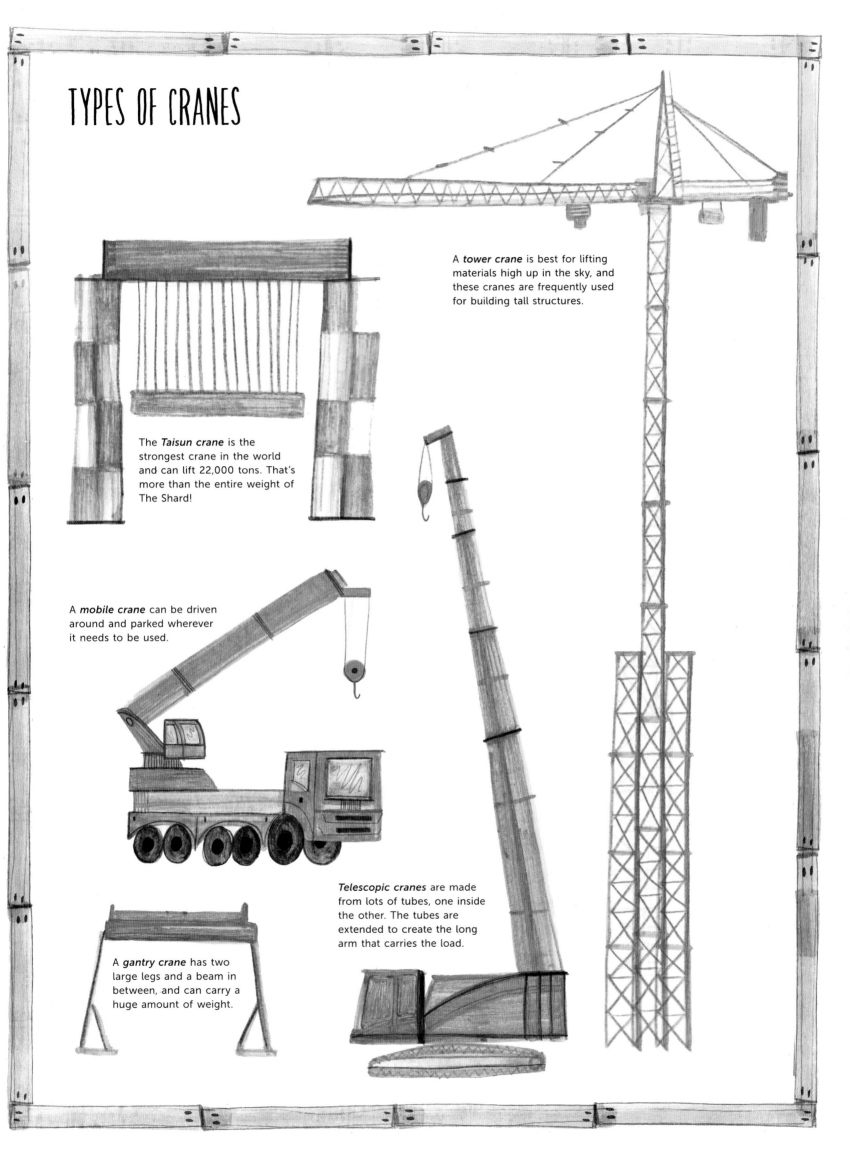

A **tower crane** is best for lifting materials high up in the sky, and these cranes are frequently used for building tall structures.

The **Taisun crane** is the strongest crane in the world and can lift 22,000 tons. That's more than the entire weight of The Shard!

A **mobile crane** can be driven around and parked wherever it needs to be used.

Telescopic cranes are made from lots of tubes, one inside the other. The tubes are extended to create the long arm that carries the load.

A **gantry crane** has two large legs and a beam in between, and can carry a huge amount of weight.

MORE SKY-SCRAPING STRUCTURES

One World Trade Center, New York, USA:
1,776 feet. Completed 2014. Its height recalls the year the US Declaration of Independence was signed, 1776.

Shanghai World Financial Center, Shanghai, China:
1,622 feet. Completed 2008. Its unusual design reduces wind pressure.

Empire State Building, New York, USA:
1,454 feet. Completed 1931. It was the world's tallest building for 40 years.

Turning Torso, Malmö, Sweden:
623 feet. Completed 2005. It's the tallest tower in Scandinavia, with a distinctive shape that makes it look like someone twisting.

Taipei 101, Taipei, Taiwan:
1,667 feet. Completed 2004. It was constructed to withstand typhoon winds and earthquake tremors.

Abraj Al Bait's Makkah Royal Clock Tower, Mecca, Saudi Arabia: 1,972 feet. Completed 2012. It has the world's largest clock faces.

The Petronas Towers, Kuala Lumpur, Malaysia: 1,483 feet. Completed 1998. The tallest twin building in the world.

Torre Mayor, Mexico City, Mexico: 738 feet. Completed 2003. It has a special skeleton, with pistons that help reduce its sway during earthquakes.

19

HOW TO BUILD LONG
THE BROOKLYN BRIDGE

The Brooklyn Bridge is a beautiful suspension bridge that was opened in 1883 to connect the boroughs of Manhattan and Brooklyn in New York, USA.

It was the first suspension bridge to use steel wires and also the first crossing over the East River. A German engineer living in the USA called John Roebling was asked to design the bridge. Very sadly, he died before construction began. His son, Washington Roebling, took over with the help of his wife, Emily Warren Roebling.

Building bridges

Bridges can be hard to build because they usually need to cross difficult terrain. When they are built over deep valleys, engineers usually put in foundations just at the ends, and make sure the bridge deck can cross without additional support. The foundations are usually made from concrete and anchored into the hillside on either side. Then a crane lifts in the rest of the bridge, whole or in sections that can be connected together.

Sometimes, if the bridge is very long and the valley isn't too deep, engineers put in more supports for the deck. The foundations are built into the riverbed, seabed, or dry valley below. If the deck is supported by cables, as it is with the Brooklyn Bridge, the towers at each end are built first, before the cables are strung across them. Then workers can build the deck and hang it from the cables above.

Caissons

Machines called caissons were an exciting new technology in the 1800s. They were used to build foundations underwater. The Brooklyn Bridge has two tall towers, and these needed strong foundations under the East River. Caissons are large airtight rooms, but they don't have a floor. As they are lowered into the water, the air inside pushes out the water, and when the walls sink into the mud of the riverbed the room is sealed. Then people can climb down into them through shafts that start above the river to build the foundations.

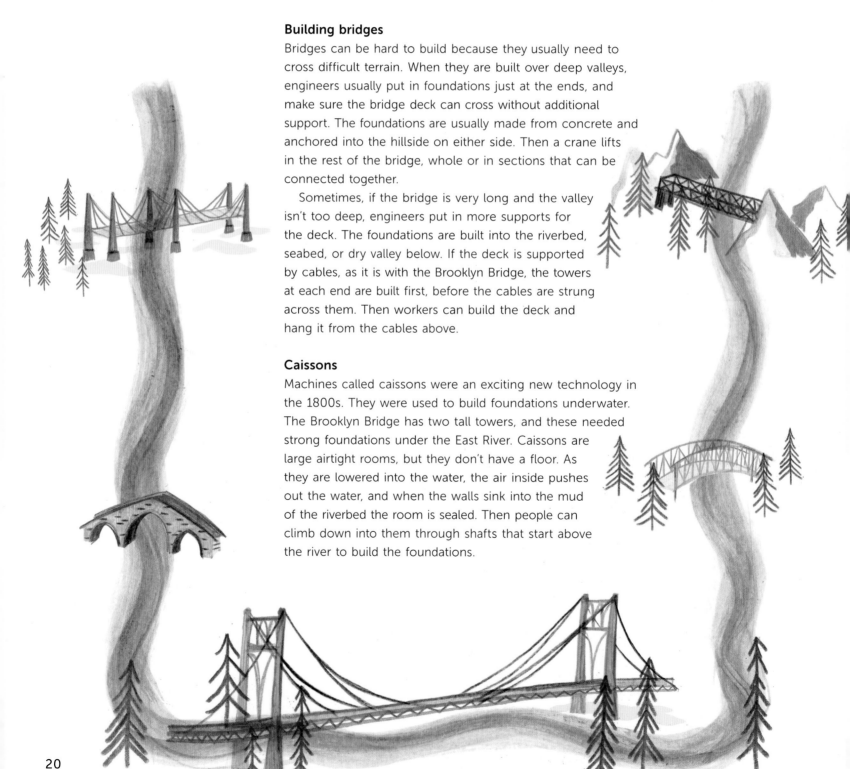

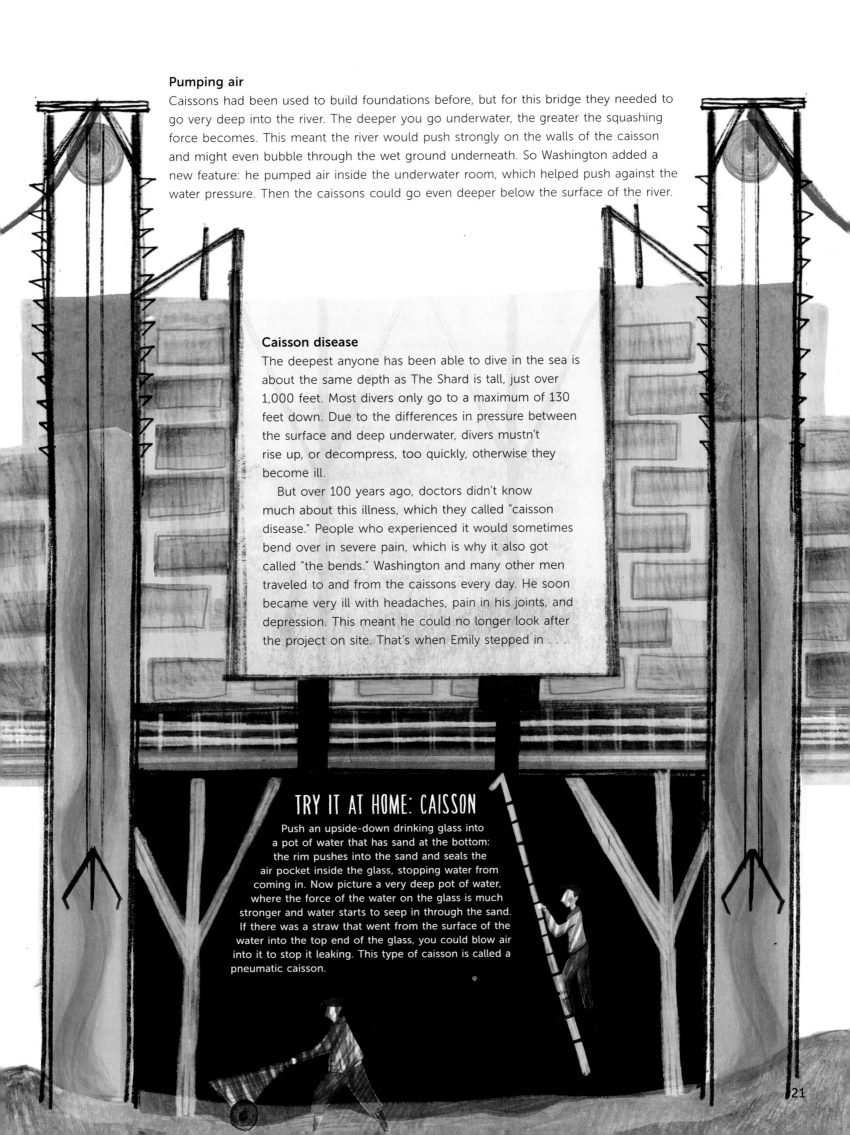

Pumping air

Caissons had been used to build foundations before, but for this bridge they needed to go very deep into the river. The deeper you go underwater, the greater the squashing force becomes. This meant the river would push strongly on the walls of the caisson and might even bubble through the wet ground underneath. So Washington added a new feature: he pumped air inside the underwater room, which helped push against the water pressure. Then the caissons could go even deeper below the surface of the river.

Caisson disease

The deepest anyone has been able to dive in the sea is about the same depth as The Shard is tall, just over 1,000 feet. Most divers only go to a maximum of 130 feet down. Due to the differences in pressure between the surface and deep underwater, divers mustn't rise up, or decompress, too quickly, otherwise they become ill.

But over 100 years ago, doctors didn't know much about this illness, which they called "caisson disease." People who experienced it would sometimes bend over in severe pain, which is why it also got called "the bends." Washington and many other men traveled to and from the caissons every day. He soon became very ill with headaches, pain in his joints, and depression. This meant he could no longer look after the project on site. That's when Emily stepped in . . .

TRY IT AT HOME: CAISSON

Push an upside-down drinking glass into a pot of water that has sand at the bottom: the rim pushes into the sand and seals the air pocket inside the glass, stopping water from coming in. Now picture a very deep pot of water, where the force of the water on the glass is much stronger and water starts to seep in through the sand. If there was a straw that went from the surface of the water into the top end of the glass, you could blow air into it to stop it leaking. This type of caisson is called a pneumatic caisson.

THE BROOKLYN BRIDGE

Each caisson used during construction was about 170 feet wide and 100 feet long. A doctor was hired to supervise the condition of the men working in them

The Brooklyn Bridge was the longest bridge in the world when it opened. It was also the first suspension bridge to use steel wire

There is a bronze plaque on one of the towers dedicated to Emily, her husband, Washington, and his father, John, the "Builders of the Bridge"

The deck is held up by steel cables that are attached to two tall towers. The four strongest cables have a diameter of almost 16 inches

The towers are made from limestone, granite, and cement and reach a height of 272 feet—which would have made them stand out against the New York skyline in the 1800s

The bridge stretches 6,000 feet across the East River, with 1,600 feet between the two towers

FRANKEIS

The bridge was officially opened on May 24, 1883 by US President Chester Arthur

23

EMILY WARREN: EXTRAORDINARY ENGINEER

Emily was born in 1843 and was one of twelve children. She had a close relationship with her oldest brother, Gouverneur K. Warren. He knew that Emily had a keen interest in science and made sure she was well educated. This was very unusual for young girls at the time, who were often not even allowed to go to school.

When her brother went to fight in the American Civil War, Emily went to visit him and that's where she met Washington Roebling, one of the other soldiers. They fell in love and exchanged hundreds of letters.

Emily and Washington got married, and after the war she visited Germany with him to study caissons.

Washington Roebling

Emily Warren

Emily's father-in-law, John Roebling, started the design of the Brooklyn Bridge but died after an accident at the dock during the planning stages. Washington then managed the project until he became ill with caisson disease.

Emily was very worried that she would also lose her husband. She took notes from Washington to preserve his knowledge about the bridge. Then she started answering his letters and helping him with his work.

In the mid-1800s, it was considered extraordinary for women to learn about engineering, but that didn't stop Emily. She studied complex mathematics, cables, and construction anyway.

Soon, Emily was supervising the work and visiting the construction site— a woman on site was unheard of at that time.

It was a difficult construction, and one which was trying new techniques and materials. Emily was soon faced with challenges.

More and more money was needed to fund it.

Workers were dying in accidents on site.

Then, in 1879, the Tay Bridge in Scotland collapsed and people worried that a similar disaster might strike the Brooklyn Bridge. The mayor of Brooklyn wanted to remove Washington Roebling from the project.

The project was even taken to court in an effort to stop it, and the Roebling family was accused of taking bribes.

But Emily slowly solved the design problems. She was very good at working with all types of people and spoke to the workers and politicians to reassure them. She convinced everyone to allow the Roebling family to continue their work.

Emily spent 11 years managing the project and must have been very proud when the bridge was finally opened. She was the first person to officially cross the bridge before the opening ceremony. She rode in a horse-drawn carriage and it's said she carried a rooster with her for good luck.

HOW TO BUILD A DOME
THE PANTHEON

The Pantheon in Rome, Italy, is an ancient monument that has been used in many ways during its long history. It has been a temple to Roman gods, a Christian church, and a tomb. It is now, incredibly, almost 2,000 years old.

When you walk inside, you notice the beautiful dome above you, with square patterns on its curves and a circular opening, or oculus, in the center, which lets in light and rain. The dome is gray and is made from concrete—a very important material for building. After water, it is the most used substance on Earth.

Concrete

If you crush a rock into powder and mix it with water, nothing much will happen. But if you burn materials like limestone first, then crush them and add water, a chemical process called hydration begins. The mixture starts to thicken up, and the liquid changes into a jelly and then into a solid.

Humans have been experimenting with crushing and mixing rocks, powders, and water for many thousands of years. The Romans burned limestone, crushing it into a powder, then mixing the powder with small pieces of rocks, broken tiles, brick, and, of course, water. The dome of the Pantheon is made from this type of Roman concrete. We still make concrete in a similar way today. The burned rock powder is called cement and the pieces of stone are called aggregates.

So: Cement + Aggregate + Water = Concrete.

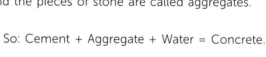

Mixing it well

To make good concrete, the cement, water, and aggregates need to be carefully measured out. Too much cement means not all of it will react with the water, so the concrete will be weak. Too much water thins the mixture out. The aggregates need to be mixed in well, otherwise the larger, heavier pieces sink to the bottom and the concrete won't be uniform and strong.

That's why we have giant drums that rotate at the back of a concrete truck. Engineers put in the correct amount of the different materials and the drum goes round and round, mixing up the concrete, so when it arrives at a construction site, it can be poured through large hoses into the right place.

A fussy material

We build all types of structures with concrete, from skyscrapers and bridges to tunnels and roads. Concrete is a very strong material when the forces acting on it are squashing it (this is called compression, remember?). If 80 elephants stood on a single concrete brick, it wouldn't crush. It also lasts for a very long time and can be buried in the ground for hundreds of years, which is why most of our foundations are built from concrete.

But concrete is weak when forces are pulling it apart—those are the forces we call tension. When a concrete structure is in tension, tiny cracks appear which can get bigger and make the structure weak. The way to deal with this nowadays is to strengthen, or reinforce, the concrete with another material, usually steel. We'll find out more about this later.

When the Romans built the Pantheon, they realized that concrete was the perfect material to use. To understand why, you need to know more about the arch.

The arch

You can make a simple arch at home with a long strip of cardstock and two erasers. Place the erasers on a table about a hand's width apart, curve the cardstock into the shape of a rainbow, and place the ends next to the erasers. If you gently push down on the top of the arch, you can imagine the forces traveling down its sides in compression.

That's what's magic about arches: they're always in compression when the force is pushing down evenly across the arch. That's why concrete works so well: it is not being pulled apart and so can resist a huge amount of weight, or loading.

The Romans built enormous bridges using this shape. The Pont du Gard aqueduct in France was constructed using three levels of stone arches stacked on top of each other to create a bridge that water could travel across to supply the nearby city.

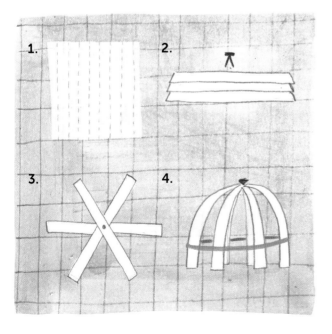

Arch to dome

A dome is an arch in three dimensions. It's a more complicated shape, and the forces travel differently through a dome compared to an arch. To make a dome, you need one piece of cardstock cut into long thin strips of equal length. Stack the strips on top of one another and stick a split pin through the center. You then need to fan the strips out and secure them with a rubber band around the base. You can see that most of the dome is in compression, but you also need tension to hold the strips together (the rubber band holding the base of the dome together).

Concrete is a great material to use to build a dome, but you do get some pulling forces near the bottom. To make sure the Pantheon was strong enough to resist this tension, the Romans built the base of the dome so it is five times thicker than its top.

THE PANTHEON

The Pantheon is the largest unreinforced concrete dome in the world and is almost 2,000 years old!

The concrete at the base of the dome is around 20 feet thick.

There are 16 Corinthian columns supporting the entrance to the Pantheon. They were carried from Egypt, down the Nile River, across the Mediterranean Sea, and finally up the Tiber River in Rome.

The concrete at the top of the dome is 4 feet thick.

The oculus at the top of the dome is the only source of light for this building.

The diameter of the dome is 142 feet, which is exactly the same as its height.

The squares all around the inside of the dome might be there to allow the concrete to keep cool and avoid cracking.

The dome of the Pantheon looks shallower from the outside because extra concrete was added to the base of the dome.

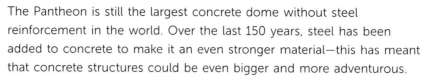

ADVANCES IN CONCRETE

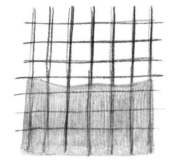

The Pantheon is still the largest concrete dome without steel reinforcement in the world. Over the last 150 years, steel has been added to concrete to make it an even stronger material—this has meant that concrete structures could be even bigger and more adventurous.

Joseph Monier

Joseph was a French gardener who lived about 150 years ago. Fed up with his clay plant pots constantly cracking, he tried using concrete instead—but even these didn't work. Then he tried putting a metal mesh inside the walls of the concrete pots, and this time they stayed intact! Joseph realized that the metal tied the concrete together and stopped it from cracking too much. Metals don't break easily when pulled apart by tension, so adding iron or steel to concrete creates a very sturdy material.

Reinforced concrete

An ancient example of mixing two materials together this way is adobe, which is mud mixed with straw. Adobe was used to make walls by the Berber people in Morocco, the Egyptians, Babylonians, and Native Americans. The straw acts like the steel—binding together the mud so it doesn't crumble. In fact, during the Victorian era, the plaster used to coat walls often had horse hair mixed in for the same reason!

Once engineers realized how well steel and concrete worked together, they started to use it in structures all around the world. Most of the concrete buildings, bridges, dams, and tunnels that you see today have a steel mesh embedded inside.

EXTRAORDINARY CONCRETE STRUCTURES

Museo de Sitio Julio C Tello, near Pisco, Peru
This museum near an ancient burial site is made from concrete which has been tinted red to match the desert around it.

La Tallera Gallery, Cuernavaca, Mexico
A beautiful concrete lattice wraps this art studio. Intricate patterns of shadows are formed when the sun shines on the walls.

MAXXI, Rome, Italy
This pale concrete structure celebrates the Romans' use of concrete over thousands of years and houses a national museum of contemporary art and architecture.

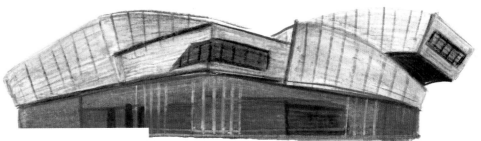

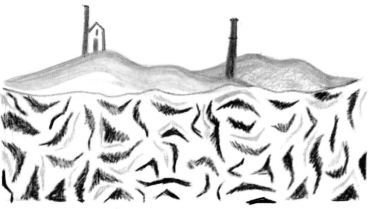

Crushed Wall, near Redruth, UK
Made using special molds, this wall looks as if it is made from fabric! It is one of the art installations on display at the Heartlands visitor attraction in Cornwall.

Lotus Temple, New Delhi, India
A temple for the Bahá'í religion, it has 27 petals surrounding a large central space. It is one of the most visited buildings in the world.

Unité d'Habitation, Marseille, France
This large residential apartment block was designed by the famous architect Le Corbusier in the 1950s.

Bank of London and South America, Buenos Aires, Argentina
A unique building that has a lot of concrete visible from the outside, this looks almost like it's made from bones.

Portuguese National Pavilion, Lisbon, Portugal
This structure really shows off how special a material reinforced concrete is! Its roof looks like a thin cloth suspended between two buildings.

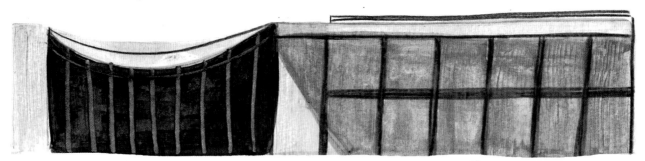

HOW TO BUILD CLEAN
LONDON'S SEWERS

All of us pee and poop every day. We each produce about 320 pounds of feces every year, and all this waste needs to go somewhere.

Take a big city like London, UK. Nearly nine million people live there, which means that every year more than TWO BILLION pounds of poop are flushed down toilets! So where does it go? And what happened before we had toilets?

Old London

Hundreds of years ago, the city of London and surrounding areas received lots of clean water from the River Thames and the smaller streams and rivers that fed into it. People didn't have flushing toilets then, so they would throw all their waste into the river. As the population increased, it became disgusting. The river was also used to dispose of human bodies and animal carcasses, which filled the water with disease-causing bacteria.

By the 1800s, there were 200,000 cesspits in the city. These were pits in the ground outside houses where people would empty their chamber pots (the bowls they used as toilets). But these cesspits leaked and caused even more pollution. Workers called "gong farmers" cleared the pits out, dumping their contents into fields and the river.

The worst thing was that the River Thames was still the main source of water for Londoners—for washing, cleaning, and drinking.

Cholera

At the same time, tens of thousands of people were falling sick with a horrible disease called cholera. Doctors couldn't understand what was causing it. Many believed it was carried by a type of polluted air, called miasma. John Snow, a doctor working in London's Soho district during a cholera outbreak in 1854, studied the patterns of people getting sick in the area and thought that the dirty water from a pump they were using was the source. But it took a while for people to be convinced.

Florence Nightingale, the founder of modern nursing, also thought that the source of disease was bad air. However, she believed in the benefits of good hygiene and that the key to a healthy life in cities was clean water and sewers. That may seem obvious to us now, but it wasn't at the time she was alive!

Florence Nightingale

The Great Stink

The summer of 1858 was unusually hot for London. The cesspits festered and the river stank even more than usual. Some people even soaked their curtains in chemicals to try to cover the stench. The politicians working in the Houses of Parliament became so fed up with the smell that they planned to leave the city. Over the years, many engineers had suggested ways to stop the waste from being dumped into the river, but nothing had been done about it.

Finally, in 1858, as the Great Stink took hold, the government asked an engineer to design a new system for London's waste.

Joseph Bazalgette

Joseph Bazalgette

Joseph was a civil engineer born on the outskirts of London in 1819.

Several tributaries flowed into the River Thames in London. And because everyone was throwing so much waste into them, they were basically sewers. Joseph decided to create brick tunnels to hide them away. Instead of allowing the polluted water to flow into the main river, he wanted to collect it and take it away from the center of London.

He did this by building three large sewers north of the river and two large ones to the south. These large sewers ran west to east.

Where the sewers and tributaries crossed, Joseph constructed overflows: these blocked the dirty water flowing into the river and helped it find a path into his new sewers.

The sewers

The Thames is a tidal river, which means the direction of its flow changes through the day. Sometimes the water flows toward the sea in the east and other times the sea levels rise and push water back into the river westward. The new sewers stopped most of the waste from going into the river in the center of London.

But it had to go somewhere, so Joseph carried it to the edge of London. Here, the tanks that had collected the waste released their contents into the river. When the sea was at low tide, the river flowed away from London and the waste was carried away with it. This meant that the Thames became much cleaner in the most populated areas of the city.

Engineers and workers had to dig up many of London's streets in order to build the new sewage system. This huge project was finally completed in 1875. By then, the cholera outbreaks had more or less stopped, in part thanks to the work of Joseph Bazalgette.

THAMES

LONDON'S SEWERS

Even today, London depends on the sewage system built more than 100 years ago by Bazalgette!

After we flush our toilets, our waste moves through a network of smaller pipes that lead to one of Bazalgette's five main sewers.

These sewers slope deeper underground as they travel toward the east so the waste flows in that direction.

More than 300 million bricks were used to build the sewer network.

High-, mid-, and low-level sewers north of the river, and high- and low-level sewers south of the river collect all the waste water and rainwater before it reaches the River Thames.

In East London, there are two pumping stations: one to the north of the river at Abbey Mills, and another to the south at Crossness. The pumps lift the waste up from the sewers that are deep underground to the level of the river.

The waste is held in tanks and released into the river when it is flowing away from the city. This untreated sewage eventually ends up in the North Sea.

FURNITURE

There are around 1,200 miles of Bazalgette's sewers underground.

The pump machinery at Crossness is surrounded by gleaming brass and colorful ironwork.

Bazalgette narrowed the River Thames to make room for low-level sewers. The underground tunnels that were built inside new embankments also made room for the first underground railway!

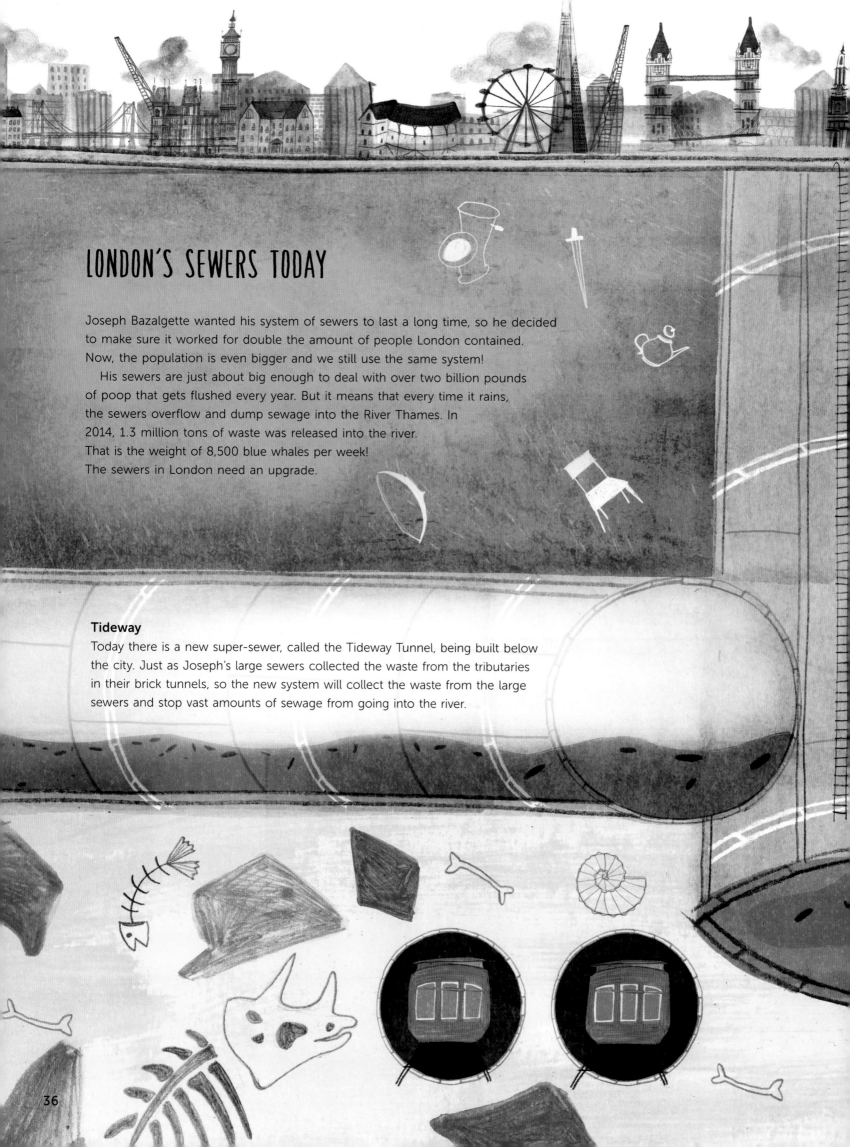

LONDON'S SEWERS TODAY

Joseph Bazalgette wanted his system of sewers to last a long time, so he decided to make sure it worked for double the amount of people London contained. Now, the population is even bigger and we still use the same system!

His sewers are just about big enough to deal with over two billion pounds of poop that gets flushed every year. But it means that every time it rains, the sewers overflow and dump sewage into the River Thames. In 2014, 1.3 million tons of waste was released into the river. That is the weight of 8,500 blue whales per week! The sewers in London need an upgrade.

Tideway

Today there is a new super-sewer, called the Tideway Tunnel, being built below the city. Just as Joseph's large sewers collected the waste from the tributaries in their brick tunnels, so the new system will collect the waste from the large sewers and stop vast amounts of sewage from going into the river.

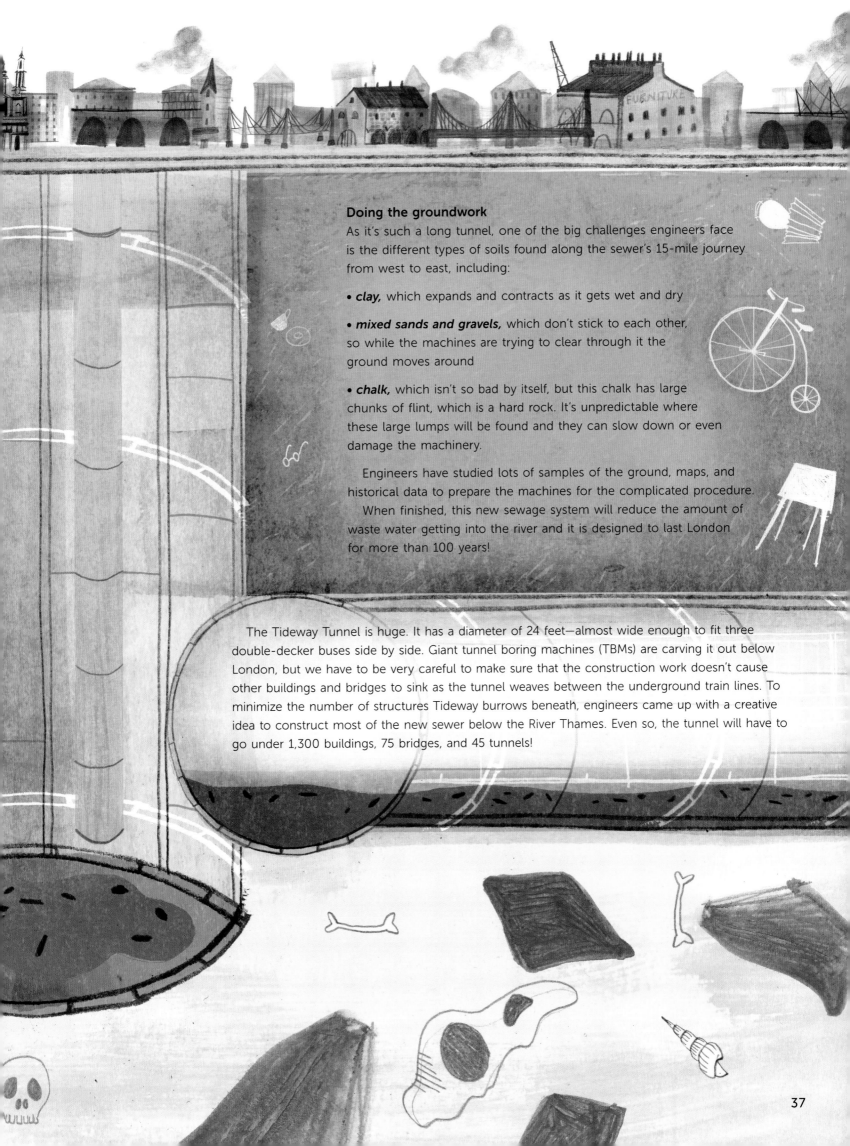

Doing the groundwork

As it's such a long tunnel, one of the big challenges engineers face is the different types of soils found along the sewer's 15-mile journey from west to east, including:

- *clay,* which expands and contracts as it gets wet and dry

- *mixed sands and gravels,* which don't stick to each other, so while the machines are trying to clear through it the ground moves around

- *chalk,* which isn't so bad by itself, but this chalk has large chunks of flint, which is a hard rock. It's unpredictable where these large lumps will be found and they can slow down or even damage the machinery.

Engineers have studied lots of samples of the ground, maps, and historical data to prepare the machines for the complicated procedure.

When finished, this new sewage system will reduce the amount of waste water getting into the river and it is designed to last London for more than 100 years!

The Tideway Tunnel is huge. It has a diameter of 24 feet—almost wide enough to fit three double-decker buses side by side. Giant tunnel boring machines (TBMs) are carving it out below London, but we have to be very careful to make sure that the construction work doesn't cause other buildings and bridges to sink as the tunnel weaves between the underground train lines. To minimize the number of structures Tideway burrows beneath, engineers came up with a creative idea to construct most of the new sewer below the River Thames. Even so, the tunnel will have to go under 1,300 buildings, 75 bridges, and 45 tunnels!

HOW TO BUILD STRONG
THE BURJ KHALIFA

Gravity pulls structures downwards, but another force—wind—pushes and pulls on structures sideways. All types of structures around the world need to be built to withstand strong winds, but some have to work harder than others to stay stable because they are more exposed.

The higher up you go, the stronger the wind is, so it's important that tall towers are designed carefully. Skyscrapers use lots of different features to stay strong.

A tower's backbone

In a storm, we see trees swaying but they rarely fall over. This is because they have deep roots that tie them to the ground and a sturdy trunk that moves but doesn't snap. Tall buildings work in a similar way. The foundations tie the structure to the ground, and we give most towers a trunk, or backbone, called its core. The core is an arrangement of walls made from steel or concrete that gives the building stability. It usually runs up the middle of a structure. When wind blows against a building, the core will sway but the foundations keep it rooted to the ground.

Pendulums

If you tie a stone to the end of a string and let it swing, you've made a pendulum. The ball in Taipei 101 stays stationary when the weather is good, but if there is a storm or earthquake it swings.

Engineers made sure that the ball was the right weight so that it swings in the opposite direction from the tower. When the building sways to the right, the pendulum moves to the left. When the building sways to the left, the pendulum moves to the right. This means the forces cancel each other out, and even though the tower still sways you can't feel it.

What if it sways too much?

It would be scary if you were standing at the top of a skyscraper and felt it moving. Engineers have to make sure that we can't really feel the movement. The Taipei 101 tower in Taiwan, which is 1,667 feet tall, has a clever piece of engineering so it's safe even in typhoons and earthquakes. It takes the form of a giant ball, measuring about 10 feet across, hanging down between the 92nd and 87th stories . . .

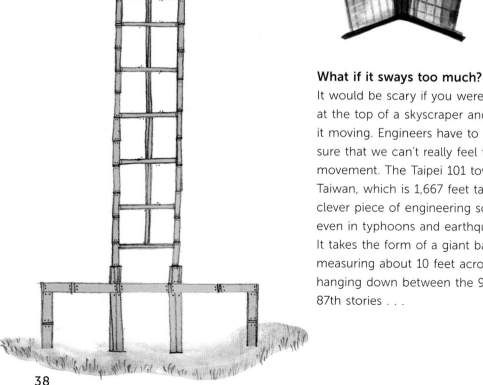

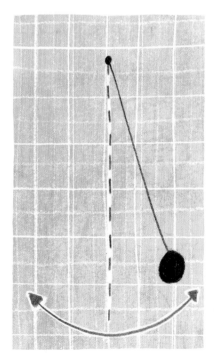

Diagrids

Sometimes a core is not enough to keep the tallest towers stable. For these, a system called a diagrid, or exoskeleton, is used. Humans have bones inside our bodies that keep us stable, rather like a core. And turtles have a shell on the outside, like towers with diagrids!

Buildings that use diagrids for stability act a bit like a hollow cylinder. They have a strong frame made of steel beams on the outside of the tower that absorbs the wind forces. This system is often called a "tubular system" because, like a hollow tube, the outside "skin" of the structure gives it strength—although the shape doesn't have to be like a tube. Diagrids are used in 30 St Mary Axe (known as "the Gherkin") in London and in the tower at Poly International Plaza in Beijing, China. The advantage of the diagrid is that you need less material than skyscrapers with cores of the same height, and you can build much taller than you could with just a core.

A bundle of tubes

A building like the Gherkin is one tube, but what if you bundle several of these together? The Willis Tower in Chicago, USA, is formed with lots of square tubes. If you look carefully, you'll see that this building has many steps as it gets taller: this is where different tubes stop. Since the tubes are connected to each other, and each one supports the other, the tower remains stable despite being so tall.

Fazlur Khan

The engineer who invented these tubular types of skyscrapers is called Fazlur Khan. He grew up in what is now Dhaka, Bangladesh, studied engineering at Dhaka University, and then moved to the USA in 1952. One of his most famous buildings is the John Hancock Center in Chicago. On its rectangular sides, you can see large Xs, one on top of the other, which form the tube that protects the building against the wind.

Fazlur Khan

TRY IT AT HOME: TUBULAR SYSTEMS

You can make your own version of the Willis Tower with straws. Take nine straws and cut them to different heights. Hold the tallest one in the center, then add slightly shorter ones around it. Wrap them with some rubber bands. The tops of the straws are like the steps in the real building. Each straw is a tubular tower and when you tie them all together, you get a very strong stability system.

THE BURJ KHALIFA

The spire is made from more than 4,400 tons of structural steel and was built inside before being hoisted up.

The Burj Khalifa in Dubai is the tallest tower in the world, standing at over 2,722 feet.

Its stability system is called a buttressed core, which is a core shaped like a tripod.

Almost 25,000 hand-cut glass panels were used for the outer layer of the towers and are designed to withstand Dubai's extreme summer heat, where temperatures can exceed a whopping 106°F.

It has the world's highest restaurant.

More than forty wind tunnel tests were carried out to see the effects of wind on the tower.

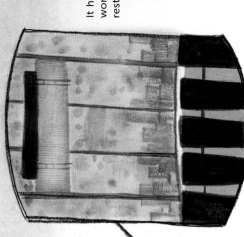

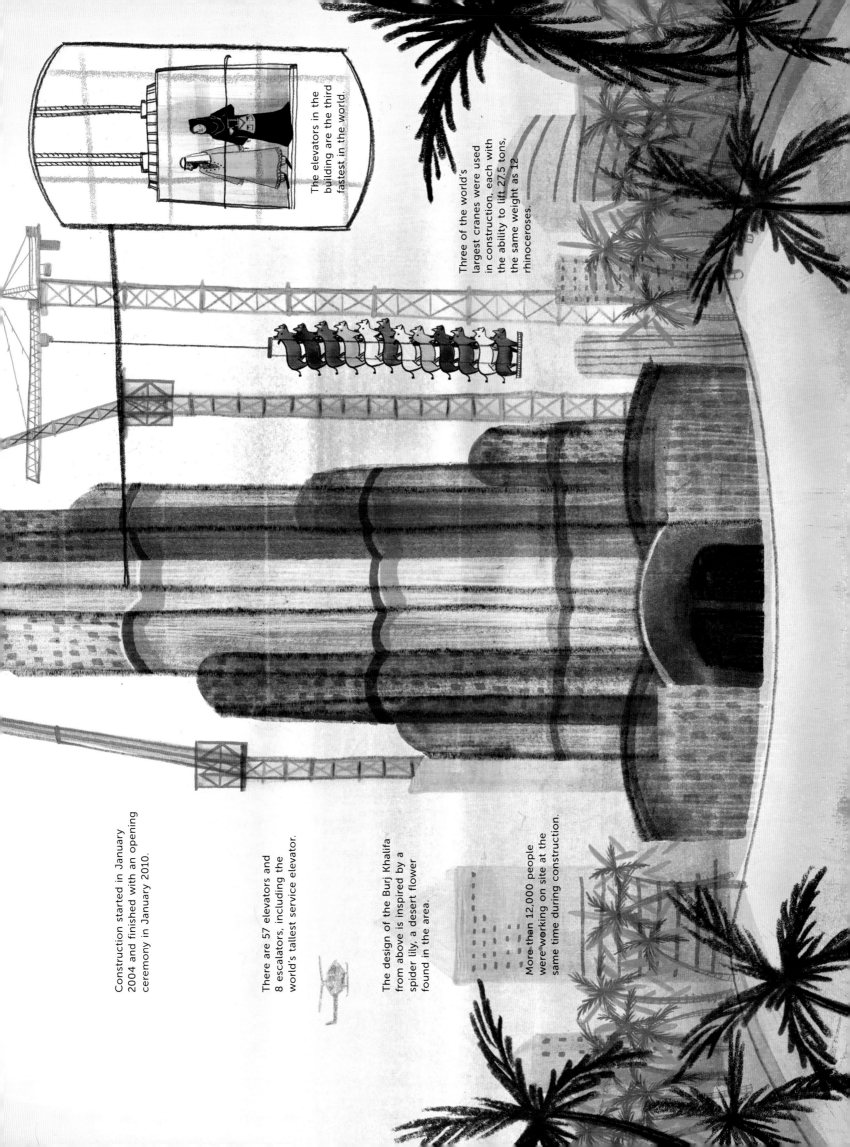

The elevators in the building are the third fastest in the world.

Three of the world's largest cranes were used in construction, each with the ability to lift 27.5 tons, the same weight as 12 rhinoceroses.

Construction started in January 2004 and finished with an opening ceremony in January 2010.

There are 57 elevators and 8 escalators, including the world's tallest service elevator.

The design of the Burj Khalifa from above is inspired by a spider lily, a desert flower found in the area.

More than 12,000 people were working on site at the same time during construction.

WHAT ELSE DOES A SKYSCRAPER NEED?

Elevators

The ancient Romans built the first apartment blocks, which were up to eight or even ten stories high. One reason they didn't go any higher was because people had to walk up and down the stairs to get to and from the topmost floors, and ten stories is a long way to climb. Even though they had the engineering to build taller, it wasn't practical for the people living inside.

In the 1800s, an inventor and mechanic named Elisha Otis was working in a factory in New York, USA, and having to hoist heavy materials from one level to another using just his muscles. He was tired of this and wanted to design a better method. By combining a platform with springs, rope, and guide rails, he created a new, safe lift. If the rope holding up the lift snapped, the springs changed shape and made the mechanism cling to the guide rails. This stopped the lift from plummeting.

Otis set up his own company to sell this new product and, in 1857, the first Otis elevator was installed in a five-story shop in Manhattan. His company lives on today; in fact, it supplied the elevators in the Burj Khalifa.

Glass

Most skyscrapers around the world have a glass skin, or facade, that protects the inside from wind and rain and lets light through. This glass is flat and smooth. If it wasn't, the reflections seen in the glass would look strange and crooked.

Before the 1950s, glass was made by rolling it while it was very hot and in liquid form. Once it cooled down, it needed to be polished. This was an expensive process. But in 1952 Sir Alastair Pilkington invented the float method of glass-making: a ribbon of glass moves out of the melting furnace and then floats on top of melted tin. The surface of the liquid tin is flat, and the glass on top is kept hot enough so it spreads evenly over the tin. Then the glass is cooled down and rolled off the tin. Glass made this way doesn't need to be polished.

TRY IT AT HOME: FLOAT GLASS METHOD

Use a baking dish and pour some vinegar into it: this represents the tin in the float glass process. Then, carefully pour some oil over the vinegar. The oil floats on top of the vinegar in a thin, even layer. This is how flat sheets of glass are made!

HOW TO BUILD ACROSS
TYPES OF BRIDGES

Bridges are built to join two parts of land together where they are separated by a deep valley, a river, or the sea.

The type of bridge engineers choose to build depends on where it is being built, how long it needs to be, what materials can be used, if it's just for people or for cars and trains, and how it will be built. There are many different types of bridges to choose from.

Cable-stayed bridge

These bridges use cables to hold up their deck. The cables go directly from the support column to the deck. The Millau Viaduct in France is a beautiful example of this bridge. It is the tallest bridge in the world, even taller than the Eiffel Tower. The first project I ever worked on, the Northumbria University Footbridge in Newcastle-Upon-Tyne, UK, is also a cable-stayed bridge.

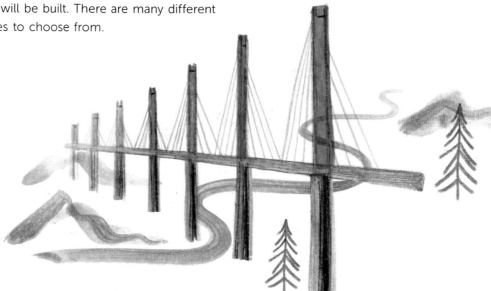

Arch bridge

These are curved bridges usually made from bricks, stones, concrete, or steel. The ancient Romans were the first to build this type of bridge in large numbers. The arch, as you will know from reading about the Pantheon, is a strong shape. Because of its curve, the forces squash it together and keep it stable. Roman engineers built spectacular aqueducts, which were tall arch bridges used for transporting water from rivers and lakes to cities.

Cantilever bridge

A cantilever is a type of structure that is supported at one end, while the other end is free, like a diving board. Imagine two diving boards face-to-face with a connection in the middle—that's how a cantilever bridge works. These clever bridges can cross long distances and are usually easier to build than bridges with cables. A crane can lift the cantilevers into position quickly. The Howrah Bridge over the Hooghly River in Kolkata, India, is an example of this type of bridge. It was opened in 1943 and carries around 100,000 vehicles every day.

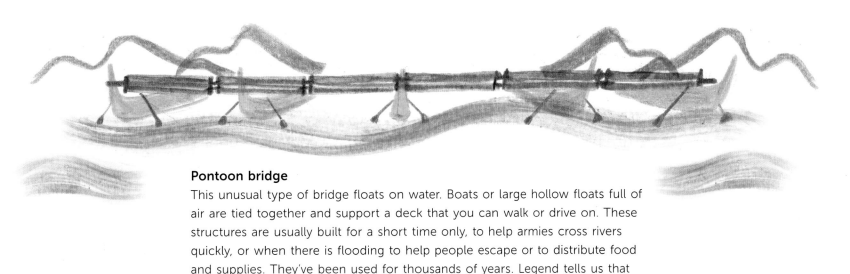

Pontoon bridge

This unusual type of bridge floats on water. Boats or large hollow floats full of air are tied together and support a deck that you can walk or drive on. These structures are usually built for a short time only, to help armies cross rivers quickly, or when there is flooding to help people escape or to distribute food and supplies. They've been used for thousands of years. Legend tells us that Emperor Xerxes, who ruled Persia in the Middle East more than 2,500 years ago, built an immense pontoon bridge to wage war against the ancient Greeks.

Truss bridge

Sometimes, the span of the bridge is so long that a beam would be too heavy and expensive. Instead, you can use a truss. A truss is a frame made up of lots of smaller beams and columns—it's much lighter and can carry heavy loads. Trusses can be different shapes: long rectangles or arches. The Südbrücke Mainz Bridge in Germany is a K-truss. If you look at its sides, you can see a pattern of Ks repeated along its length, facing different ways depending on how the forces flow.

Trestle bridge

Trestle bridges are made from wood or, sometimes, steel. They have lots of columns or supports (made like trusses) one after the other to support the deck. Many were built in the 1800s and early 1900s to carry trains. They are quite common in the USA. Rollercoasters are often designed like trestle bridges because it means the forces from the fast-moving cars can be safely absorbed by the many supports, and it's an easy way to build really high for those scary drops!

HOW TO BUILD STABLE
TE MATAU Ā POHE

The Māori are the indigenous people of New Zealand, who are believed to have arrived there from elsewhere in the Pacific around 750 years ago. In Polynesian mythology, Māui was a demigod— a clever, gifted, strong part-human, part-god.

The story goes that Māui's brothers had planned a fishing trip, but didn't want Māui to come along. So he crawled into the hull of their canoe at night to hide, taking with him a fishing line he had woven himself and a hook made from the jawbone of his ancestor Muri-ranga-whenua. Once they were far out at sea, Māui suddenly revealed himself. He threw his magical fish hook, or hei matau, over the side of the canoe. Suddenly, he felt it touch something and tugged gently: he had caught an enormous fish! He hauled the fish to the surface with his brothers' help, warning them they should first appease Tangaroa, the god of the sea, before cutting the fish up. But his impatient brothers began to carve it into pieces.

New Zealand

The legend says that this fish formed the North Island of New Zealand. It would have been a flat island, but because Māui's brothers cut into the fish they created mountains, valleys, and a rugged coastline. Even today, this island is called Te Ika-a-Māui, which means, "the fish of Māui."

Māui had a big influence on the other islands too. The South Island is known as Te Waka a Māui, "Māui's canoe," and Stewart Island is Te Punga o Te Waka a Māui, "the anchor stone of Māui's canoe."

A moving structure

Whangarei, a city on New Zealand's North Island, needed a new bridge across the Hātea River. The river is an important route for boat traffic, so the bridge had to be able to open quickly and safely to let tall boats coming from the harbor sail up into the city.

Architects and engineers drew from the story of Māui in their design. The hei matau is an important part of Māori culture, as the sea is a rich source of food. The hook symbolizes abundance, strength, and determination, and is considered a good luck charm by those who fish or travel by sea. A beautiful bridge inspired by the magical fish hook would perfectly reflect the local culture and heritage.

The new structure they built is called Te Matau ā Pohe, which means "the fish hook of Pohe." Pohe was a famous Māori chief who welcomed the first European settlers to the area. An island was named after him, and this bridge connects the city of Whangarei to Pohe Island.

Center of gravity

Every object has a center of gravity: this is the point around which its weight is even on all sides. For an evenly shaped object such as a ball or book, the center of gravity is in the middle. But for uneven shapes, such as our bodies, the center of gravity isn't in the middle. Ours is actually higher than our hips, because the top half of our body is heavier than the bottom half.

When designing a moving bridge, designers need to understand where the bridge's center of gravity is so it doesn't take too much force to move it.

Rolling bascule

On the Te Matau ā Pohe bridge, one section opens. Two giant J-shaped beams that look like the fish hook make this possible. Their clever, curved shape means they can roll into the open and closed positions quickly.

If you lie on the floor on your back and lift your legs up in the air, you are shaped something like the movable J-beam on the bridge. Your center of gravity will be somewhere near your chest, which could make lifting up the deck difficult. To fix this, the engineers added lots of weight at the top of the J (this would be like attaching weights to your raised feet). Now the center of gravity is closer to the curved section (your rear), so the deck can roll easily when pushed up. This type of rolling bridge beam, which is carefully balanced by its weight, is called a bascule.

TE MATAU Ā POHE

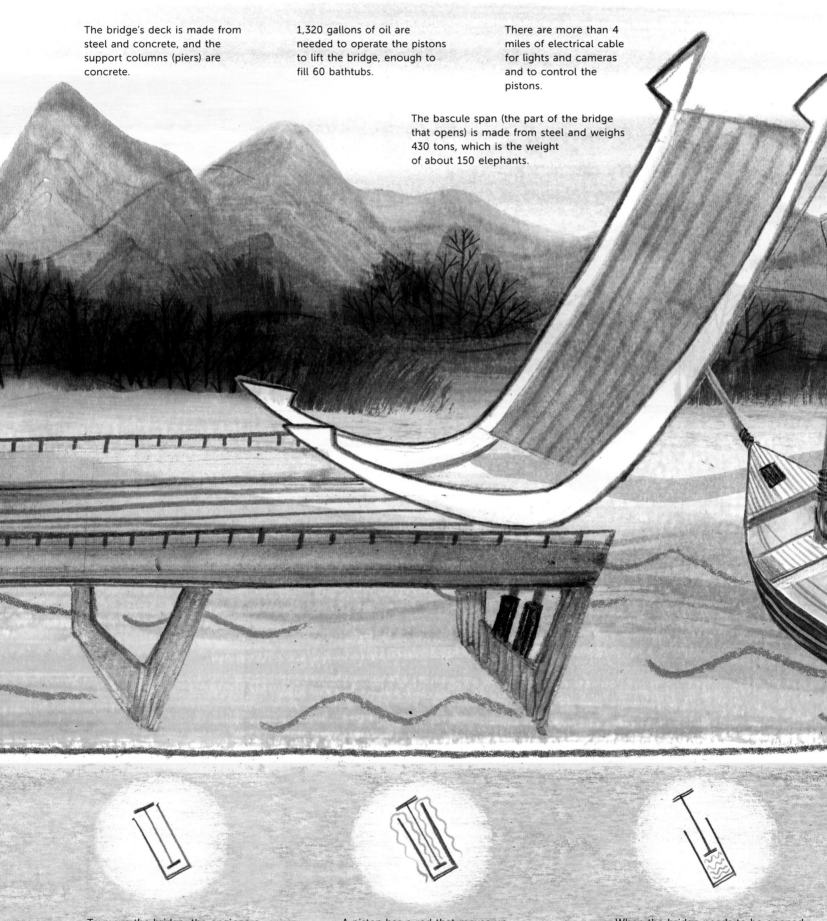

The bridge's deck is made from steel and concrete, and the support columns (piers) are concrete.

1,320 gallons of oil are needed to operate the pistons to lift the bridge, enough to fill 60 bathtubs.

There are more than 4 miles of electrical cable for lights and cameras and to control the pistons.

The bascule span (the part of the bridge that opens) is made from steel and weighs 430 tons, which is the weight of about 150 elephants.

To move the bridge, the engineers used two arms that could get longer and shorter, called pistons.

A piston has a rod that moves up and down inside a cylinder of oil.

When the bridge needs to be opened, an operator presses a button and oil is pumped into each piston.

Three Māori iwi (tribes) blessed the site before construction started and were instrumental in selecting the final name for the bridge.

When designing the piers that rise from the seabed to support the bridge, engineers had to think about the force of a ship crashing into them.

The bridge is designed to resist earthquakes. Engineers left a small gap between the moving and fixed parts of the bridge. This way, they can shake freely during an earthquake and not damage each other.

The bridge was officially opened in July 2013. It stretches across 870 feet of the Hātea River.

This makes the piston become longer.

The pistons then push up on the J-beam, and the beam rolls backward along its curved edge . . .

to open a path for tall boats.

BRIDGE SAFETY

Structural and mechanical engineers as well as architects work together to create carefully-crafted drawings of what a finished structure will look like. Engineers have to use their special knowledge to see in a completely different way and create a strong skeleton or frame so that structures can survive the pull of gravity and natural disasters.

Resonance and frequency

Think about swinging on a swing. Did you know that over a set period of time, say ten seconds, you go back and forth the same number of times whether you're swinging low or going up really high? The number of times you complete a full swing back and forth always remains the same! You can try this in the park or even model it at home with an orange tied to a string. Push the orange with different amounts of force and count how many times it passes the midpoint of its swing in ten seconds. The number of times the orange swings in one second is called its frequency.

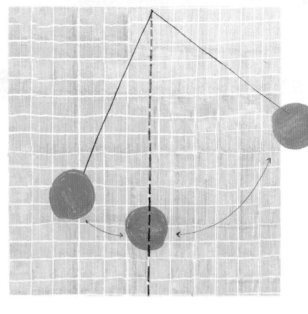

If you push the orange every time it reaches its highest point, you'll notice two things—first, it still passes the midpoint the same number of times as if you don't push it. And second, it moves further. Your hand is matching the frequency of the orange, which makes it swing a lot more. This is called resonance.

All structures also have their own frequency, the number of times a structure sways per second. When the surroundings match that frequency, creating resonance, it can be very dangerous for buildings and bridges. Engineers need to carefully calculate a structure's frequency so that it stays upright and secure. The Tacoma Narrows Bridge is an example where wind effects caused the whole structure to collapse.

The safety gap

Most moving bridges are designed to have some form of gap to allow them to get shorter and longer as the seasons change. Materials expand (get bigger) when hotter, and contract (get smaller) when colder. Usually, these gaps are sealed with rubber. But the Te Matau ā Pohe has an open gap, because of another dangerous force.

Earthquakes

New Zealand experiences over 15,000 earthquakes every year. Luckily, most of them cannot be felt because they are so tiny, but around 100 of them are strong enough for you to feel. Earthquakes shake the ground forward and backward, and up and down. Structural engineers need to design structures so they don't collapse during earthquakes.

The Te Matau ā Pohe bridge had the added complexity that part of the bridge lifted. The designers decided that it was safest to have an open gap between the western and eastern parts of the bridge, which is small enough so cars can drive over it safely! Each side of the bridge had a different natural frequency, and it was calculated that neither frequency matched the frequency of a typical earthquake, so the bridge wouldn't move dramatically if one occurred.

HOW TO BUILD WATERTIGHT
KATSE DAM

The Earth is sometimes called the Blue Planet because more than 70% of its surface is covered in water.

Humans need fresh, clean water to survive—in fact we can't live for more than three days without drinking any. Earth has plenty of fresh water in lakes and rivers. But fresh water is only a small part of the planet's total water, most of which is salty or difficult to reach. Imagine that all the water on our planet was represented by the size of a football field. The freshwater lakes would be the size of a small pillow, and the rivers we can drink from the size of a coaster I use under my glass. This means it can be difficult to source clean water. For thousands of years, we have been engineering clever ways to find, store, and move this precious resource.

Storing water

One way to store water in rivers is to block the river with a structure called a dam. By building a dam, we can create a large lake or reservoir and use the water for irrigating crops, generating electricity, or for everyday use in our homes.

There are many types of dams. Some are made from large piles of mud or clay, others from stacks of small stones. But the largest ones are usually built from concrete.

The beginnings of the Katse Dam

Lesotho is a landlocked country, surrounded on all sides by South Africa. It has abundant supplies of fresh, clean water from the Senqu River. The area around the river is hilly, which meant that the people of Lesotho could build a dam and use the water to generate electricity, or hydro-power, for themselves. South Africa would also be able to benefit, using the water to supply its large population and growing industry. So the two countries came together to create the Lesotho Highlands Water Project, one of Africa's largest water transfer systems. Part of this immense engineering project is the Katse Dam.

Double-curvature arch

The Katse Dam is located on the Malibamat'so River, which flows within a deep gorge and later joins the Senqu River. A double-curvature arch was one of the safest and cheapest options for the shape of the dam. These dams work best for narrow gorges like the one on the Malibamat'so River, and they use fewer materials to build. A dam like this is curved from top to bottom and also from side to side.

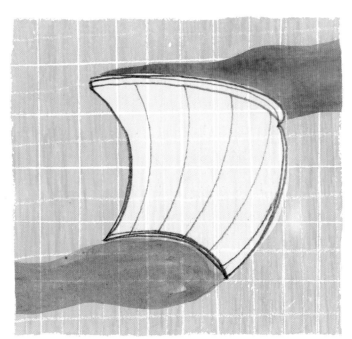

Now, remember that water is much heavier than air, and if you dive into water, the deeper you go, the more the force of it tries to squash you. You can sometimes feel this pressure in your ears if you're swimming at the bottom of a pool.

The water that dams are built in is very deep, so engineers need to calculate how large this force is and how to attach the dam at its base and sides securely to make sure nothing leaks. They also calculate how thick the base of the dam needs to be so that the water doesn't push it over.

Cool concrete

One of concrete's unique properties is that when the chemical reaction occurs between the water and cement, heat is released. A huge amount of concrete was poured to create this large dam, which meant a lot of heat was generated. If the concrete had gotten too hot, it would have cracked as it cooled down, which would have allowed water to leak through. So during the summer, when it was very hot, engineers added flaked ice into the mixer to keep the temperature below 59°F. In the winter, to stop the concrete from freezing, they used heated water to keep the concrete above 44°F.

KATSE DAM

The Katse Dam was completed in 1996. It is over 600 feet tall, the second highest in Africa and in the top ten largest concrete arch dams in the world.

The dam stretches more than 2,300 feet across.

It's almost 200 feet wide at the base and 30 feet wide at the top.

46 million cubic feet of rock were excavated, enough to fill over 500 Olympic-sized swimming pools.

Double-curvature arches are not that common even though they use less material. They are complicated to build and they need engineers with specialist knowledge.

The reservoir can hold over 70 billion cubic feet of water; that much water would fill Sydney Harbor more than three and a half times.

Trout and yellowfish can be found in the reservoir.

NATURE'S DAM ENGINEERS

Beavers are large rodents found mainly in North America, and some of the world's best animal engineers. They live in lakes and ponds, building homes called lodges to keep themselves safe from predators such as bears, wildcats, and otters. These special homes have a secret entrance underwater that other animals can't access.

Using their large front teeth, beavers gnaw through small- and medium-sized tree trunks, adding them to the dam until a deep enough pool of water is created. Then they make their home out of sticks and branches. They look after their babies here and hide all winter to keep safe.

The world's biggest beaver dam can be found in Wood Buffalo National Park, Canada. It stretches for almost 3,000 feet, longer than the Katse Dam. It's so big that it can even be seen from space! It's perfectly positioned to collect lots of water, and it's on the edge of a dense forest that provides food and plenty of wood for building.

ANCIENT WATER ENGINEERING

Dams are an excellent way to store moving water. But what if you can't easily find any water to store? Like other civilizations around 3,000 years ago, the ancient Persians struggled with low water supplies in their desert climate. People throughout the Middle East and northern Africa had to develop ingenious ways to locate fresh water and move it around. The system the Persians used was called the kariz or qanat. It consisted of a number of wells joined together by a tunnel.

Qanat

Below the Earth's surface, there are layers of rock where pools of water sit. These are called aquifers. The first step for the workers (called muqanni) who built the qanats was to find this hidden water. They started on hills and mountains, digging deep wells and lowering buckets down to check if any water appeared. If it did, they would measure how much water appeared. If there wasn't much, it wasn't worth building a qanat.

Once they found an aquifer with lots of water, they dug many wells in a line one after the other, and each one went slightly deeper than the last so that when they were finally connected, the water would flow downhill. Then they dug a tunnel at the base of the mountain where they had found the aquifer. This tunnel joined up the bottoms of all the wells until finally it connected to the very first well. Then the water would gush into the tunnel and flow out of the mountain into channels dug by the muqanni, which then carried the water to their town or village.

Long-lasting source of water

The muqanni regularly cleaned out the system by removing extra mud from the wells and tunnels to make sure the water had a clear pathway. In Iran, there are believed to be over 35,000 qanats, many of which are still being used today. One of the oldest and largest examples is in the city of Gonabad. It is 2,700 years old, and the tunnel is 28 miles long. The main well is more than 1,000 feet deep, which could comfortably fit in The Shard, and still provides water to 40,000 people today.

qanat

HOW TO BUILD UNDERGROUND
TYPES OF TUNNELS

Dom Pedro II was crowned the second-ever emperor of Brazil in 1841 when he was just 14 years old. He was especially interested in engineering, astronomy, literature, and languages, and wanted to connect his entire empire by rail. The railway would need to journey through the coastal mountains of the state of Rio de Janeiro. Seventeen years after his coronation, work finally began on a large tunnel called the Túnel Grande.

The Túnel Grande stretches an incredible 1.4 miles through the Serra do Mar mountain range in Brazil. It's almost 14 feet wide and took 7 long years to complete, yet was only one in a series of 15 tunnels that were part of the Dom Pedro II Railroad.

Prehistoric tunnelers

Scientists believe that a strange network of tunnels in Brazil was dug out by an ancient species of giant sloth! These animals worked in teams to dig out enormous burrows, called paleoburrows. The tunnels are roughly 4 feet wide, and one of these burrows has many branches that, together, are over 2,000 feet in length. Scientists can't think of any natural geological way in which these could have been formed, and have found claw marks on the walls, which is where the giant sloth theory comes from.

Rock hard

The Túnel Grande had to be cut through solid rock. The only way to do this at the time was using tools like chisels and hammers, as well as gunpowder. It is said that the rock was so hard that the heaviest of blows from the construction workers produced just a little dust. There were no mechanical drills and no dynamite, an explosive invented in 1867 by Swedish chemist and engineer Alfred Nobel, after whom the Nobel Prize is named. Dynamite made such a difference to construction that when another tunnel of the same length was excavated after its invention, the process took only 11 months.

To try to speed things up, the workers dug four vertical shafts down through the mountain. This way they could dig the tunnel out at different points at the same time. They also dug from both ends of the tunnel. Around 400 laborers worked day and night in shifts.

On June 30, 1864, when the two ends of the last stretch of the tunnel finally met, Emperor Dom Pedro II visited the site. It is said he leaned into the tunnel and threw money down to the workers!

THAMES TUNNEL

A few decades before the Túnel Grande started construction in Brazil, the first-ever tunnel under a navigable river (one deep and wide enough to carry boats) was built under the River Thames in London, UK. There had been huge problems with congestion on the few bridges crossing the Thames back in the 1800s. It could take hours to get through the horse-drawn vehicles and people clogging up the bridges, so it was time to try something new. Since there were so many ships using the Thames, adding more bridges could block them. The ideal solution was to create a way to cross under the river.

Shipworms

Marc Brunel was born in Normandy, France, in 1769. He became an engineer and traveled around the world before settling down in London. Others had tried to build a tunnel under the Thames but failed because the ground was so wet and of poor quality. But Brunel had a new idea, inspired by a worm!

The *Teredo navalis*, or shipworm, has two razor-sharp "horns" on top of its head. As it moves through wood, these horns grind it into a powder to create a tunnel. It eats the powdered wood, which travels through its digestive system. When it excretes this, the poop-paste lines the tunnel behind its rear. The air in the tunnel hardens the paste and makes the tunnel strong. Brunel decided to try to copy the shipworm in finding a way to build under the Thames.

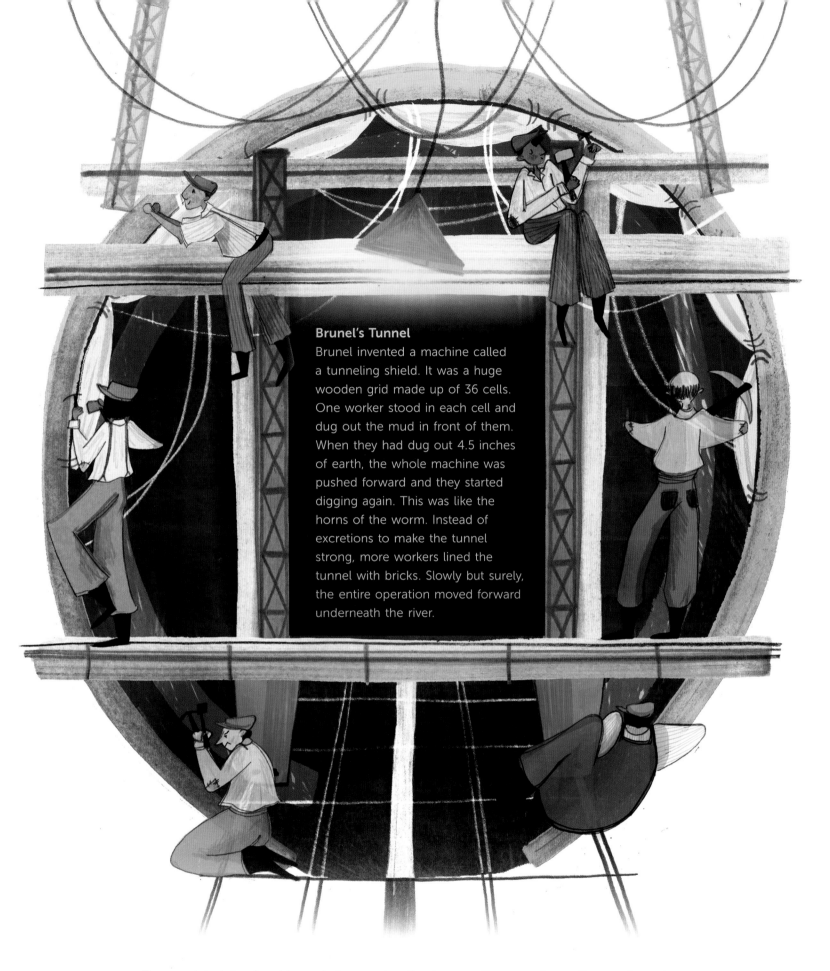

Brunel's Tunnel

Brunel invented a machine called a tunneling shield. It was a huge wooden grid made up of 36 cells. One worker stood in each cell and dug out the mud in front of them. When they had dug out 4.5 inches of earth, the whole machine was pushed forward and they started digging again. This was like the horns of the worm. Instead of excretions to make the tunnel strong, more workers lined the tunnel with bricks. Slowly but surely, the entire operation moved forward underneath the river.

They ran into lots of problems during construction because the ground sometimes collapsed and water from the river flooded in. Many people were injured and some died. The Thames Tunnel finally opened in 1843, 18 long years after work had started.

At first people walked through it, but later it was modified for trains. Today, the London Overground line between Rotherhithe and Wapping stations runs along this very tunnel.

THE HUMBLE BRICK

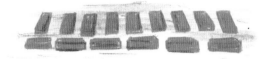

Our ancestors have been building from materials readily available around them for thousands of years: they've used trees, rock, animal skins, and even mud. Archaeologists have found evidence of mud bricks that were used around 9000 BCE in the ancient city of Jericho in the Middle East. The residents molded flat pieces of clay, baked them in the sun, and then built homes in the shape of beehives. To make bricks even stronger, engineers living in cities around the Indus Valley (in modern-day Pakistan) in 2600 BCE heated them to much higher temperatures in ovens called kilns.

Today, we "fire" the bricks at even higher temperatures, between 1,400 and 2,200°F. This makes the particles of the clay fuse together and turns it into a different material: ceramic, which is more similar to glass than dried mud. These are the strongest types of clay bricks, and each one can support the weight of five elephants!

Ingredients

Clay is a type of soil made up of volcanic rocks that have been ground down to a fine powder. Long ago, these rocks were picked up by moving water, wind, and ice. As they moved, they combined with particles of other minerals, such as quartz, mica, lime, and iron oxide. This mixture was deposited far away from its original source in layers at the bottoms of rivers, valleys, and seas. Plants and animals lived and died in these areas, adding layers of organic or living materials to the ground. Gradually, over millions of years, with changes in temperature and pressure, these layers of crushed rocks, minerals, and organic matter turned into clay.

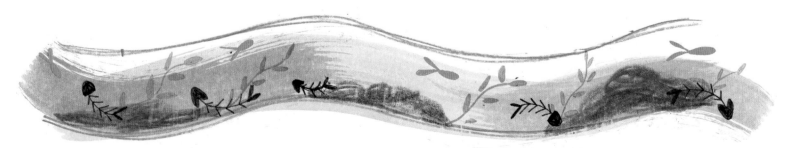

London clay

Around London, where the River Thames has deposited soils, there are deep layers of clay that are incredibly old. The upper layers, which are slightly tinted red because of the presence of iron, are "newer" clay and are around 20 million years old. Below that, the clay is blue-gray in color; it's purer and it can be up to 50 million years old!

This type of clay was mined and used to make the bricks which still hold together the Thames Tunnel over 150 years after it was built.

Bricks, bricks, bricks

Around 1.4 trillion bricks are made every year around the world. China alone manufactures 800 billion, and India makes around 140 billion. Lego makes about 45 billion bricks every year!

HOW TO BUILD MOVING THINGS
SAPPORO DOME

Bridges are among the earliest structures that we know about with moving parts. Take, for example, Old London Bridge, UK, which was completed in 1209 and demolished in 1831. It had a drawbridge, a section that could be lifted up to allow tall ships to pass through.

When canals were built in the UK in the late 1700s and early 1800s, bascule and swing bridges were built where the water passed through farmland, and these usually had to be operated by people. It took a long time to get to the point where we could build revolving restaurants and large, retractable roofs for stadiums. These have only really become popular in the past few decades, because they are complex and we need computers to help with the calculations required to build them.

Different types of engineering

Any large moving structure brings together lots of different types of engineering to make it work. Architects design what it will look like. Structural engineers look at the materials and how to keep the structure stable. Mechanical engineers consider the machines that will make the parts move, and electrical and systems engineers design the power, safety, and control systems to power the machines. Everyone has to work together to create the best and safest design.

Ingenious moving structures

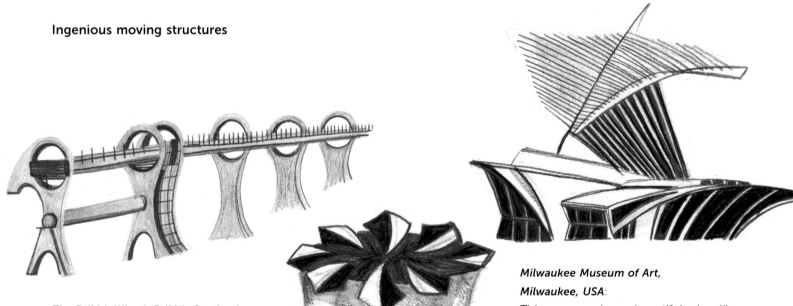

***The Falkirk Wheel**, Falkirk, Scotland:*
This is a bridge for boats! It has gondolas—watertight compartments—that you can sail into at a high level. Then the entire wheel rotates and drops you off at the lower level (or the other way around).

***Qi Zhong Stadium**, Shanghai, China:*
This stadium has a sliding roof structure which looks like an opening magnolia flower. Opening the roof only takes eight minutes!

Milwaukee Museum of Art, Milwaukee, USA:
This museum has a beautiful wing-like structure called a *brise-soleil* (French for "sun breaker") that moves according to the intensity and position of the sun. The raised wings create shade and stop the inside of the building from getting too hot, and they fold down again at night or in bad weather.

Sapporo Dome

Sapporo is Japan's northernmost major city and it experiences around 16 to 20 feet of snow during the year. It was one of the host cities for the 2002 Soccer World Cup. Baseball is also very popular in Japan, so the city wanted to construct a stadium that could accommodate both games in all seasons. This meant a large roof was needed.

The World Cup rules meant the soccer field had to be made from natural turf, but grass won't grow well under a covered roof. So engineers designed a "hovering stage" for the field, enabling it to be kept outside. When there is a soccer match inside, the entire field is moved into the dome!

Moving a field

The stadium is usually set up for baseball games, with artificial turf and seating angled around the field to give the best view for watching this sport. The hovering stage, which is 280 feet wide, 400 feet long, and weighs close to 9,000 tons, stays outside most of the time, so the fresh air and sunlight will allow natural grass to grow.

To convert it into a soccer stadium, the artificial turf is dismantled and stored away. Some of the lower seats are moved from their angled position to line up with the straight sides of the hovering stage, which slides in through huge doors that are revealed when a large section of seating at one side of the dome folds up and retracts. Once the field is in place, the entire floor is rotated by 90 degrees and the retractable seating slides back into place to complete the transformation.

But how does the stage move? Large blowers are used to lift the stage up slightly and 24 wheels enable it to roll into position—90% of the weight of the stage is carried by blowing air!

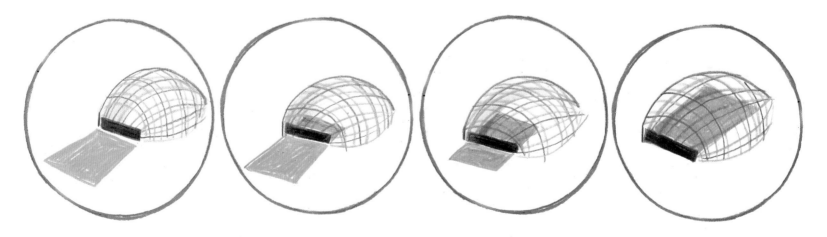

SAPPORO DOME

The large roof is made from a framework of steel trusses and cables approximately in a dome shape. This type of structure is called a shell. Shell roofs are normally closed in on all sides, but because this domed stadium needed a large opening at one end to allow the stage to move in and out, they nicknamed it a "clopen" shell (closed and opened). It is more than 200 yards wide and has to resist gravity, wind, snow, and earthquakes.

It takes five hours to convert the baseball stadium to a soccer field.

To build the roof, more than 30 temporary towers were constructed to support the beams as they were lifted in by crane.

The air-lifted wheel-drive system used to move the field was the first of its kind in the world.

Nearly 42,000 people can fit into the Sapporo Dome when it's set up for soccer matches.

There are moveable structures inside the Sapporo Dome too, including two stands that rotate, two large three-tier stands that fold and move sideways to make space for the stage to enter, and a baseball pitcher's mound that retracts underground.

Large steel beams support the floor of the hovering stage to make sure it's firm enough for the soccer players to run around on without bouncing.

A team of 10 technicians is based permanently on site to maintain all the moving equipment.

The large blowers that lift up the stage can change the amount of air they're blowing if the stage needs to be moved higher or lower to make sure it moves evenly and smoothly.

COMPUTER MODELING

Complex structures such as the Sapporo Dome are possible today because of computers. Previously, engineers could only use more basic calculation tools. They couldn't work out the weight of their structures, the variable forces such as wind and earthquakes, or how the forces would channel through the structural skeleton with as much accuracy as we can today.

Now we can create virtual 3D models of our structures, including their floors, beams, and columns, all in the right materials. Gravity, wind, and even earthquake forces are applied to these models. A computer runs millions of calculations, then tells us what forces are going where, whether our structure is strong enough, and how much it moves.

Ada Lovelace

Computers are a fairly recent invention, but computer science was born in the 1840s. Ada Lovelace developed the first algorithms in an era when women were usually denied education.

She was born in 1815 in London, UK. Her mother insisted upon a strict regime of science and mathematics learning. Ada loved designing boats and flying machines when she was a child.

When she was older, she was introduced to Charles Babbage, an inventor who was designing complicated calculating machines, including the Difference Engine and Analytical Engine. Ada was inspired and came up with a method by which the Analytical Engine might calculate a mathematical sequence of numbers called Bernoulli numbers. She also had theories about the machine's potential use in manipulating any system of symbols, including music.

Even though the Analytical Engine was never built, Ada had created what many consider to be the first-ever computer programs.

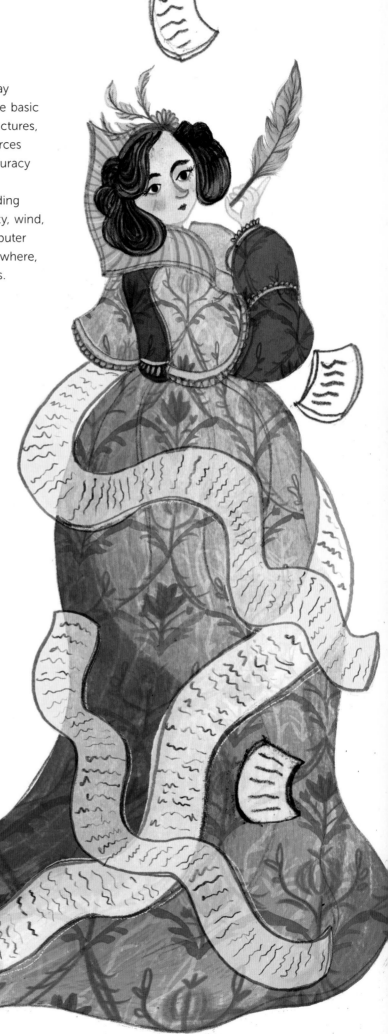

Alan Turing

Alan Turing was an English mathematician who was instrumental in breaking the Enigma code, which was used by the Germans to send top-secret messages during World War II.

He was born in 1912 and studied mathematics at the University of Cambridge. While he was studying for his PhD, he developed the idea of a "universal computing machine" that could solve complex calculations. This was called the Turing Machine and later led to the invention of digital computers. Alan also studied cryptology: the science of codes and ciphers that can be used to send secret messages.

During the war, he developed the "Bombe" with other mathematicians. It was a machine that could break German codes on a huge scale. After the war, he created a design for a machine called the Automatic Computing Engine. This worked more like the digital computers we have now and it could store programs in its memory.

Alan was an incredible mathematician and he's considered the father of theoretical computer science and artificial intelligence.

HOW TO BUILD ON ICE
HALLEY VI

Halley VI is a research station built on the coldest, driest and most remote continent in the world: Antarctica.

Antarctica is one of the most inhospitable environments on our planet, but the British Antarctic Survey has a team of scientists there who study the atmosphere, the sea, the ice, and climate change. This research facility is on the floating Brunt Ice Shelf, where temperatures rarely rise above 32°F in the summer, and during the winter, when there is no sunlight for 105 days, they drop down to nearly -67°F! Because it's so cold and the weather is so harsh, from March until November no one can actually access this location, and a smaller team remains isolated. But that's only if it's safe for them to be there! Halley VI is the sixth research station to be built on this site and was officially opened in 2013. The first four were destroyed by the build-up of snow in the harsh climate, and the fifth was in danger when it looked like huge chunks of ice near it were suddenly going to break away.

A station with legs

Building in Antarctica is very different from building on other continents because the ground is always changing. Snow can pile up to 5 feet or more every year, or an ice shelf can break apart. So a clever way to make sure the latest research center would last a long time was to design it to move.

The station is made up of eight modules—some for living and sleeping, some for generators that supply power, and others for laboratories. Each module is supported on giant skis with hydraulic legs. The hydraulic legs are powered by pistons, which means that each leg can be made longer or shorter. If it snows and the skis get buried, the pistons shorten the legs so the skis are lifted back to the surface of the snow.

Skiing to safety

To move the station, first the modules are disconnected from one another and then each one is tied up to something that looks like a cross between a tractor and an army tank. The tank drives across the snow and drags the module on its skis. Halley VI has already been moved once, about four years after it opened, because a large crack had appeared in the ice shelf. The station was shifted 14 miles away from its original site.

It was very important to protect the environment during construction, so the engineers made sure that the strict Environment Protocols (or rules) of the Antarctic Treaty were respected.

The modules had to be as light as possible so they were easy to move.

The material that surrounds the modules is designed to keep as much heat inside the structures as possible.

On-site construction could only be carried out during 10 weeks of the Antarctic summer. So although it took 4 years to complete, the actual work period was only 40 weeks.

The hole in the ozone layer, which is dangerous for our planet, was discovered at an earlier Halley station in 1985 by British scientists.

The Brunt Ice Shelf is 140 yards thick. It flows slowly into the Weddell Sea.

The modules are colorful inside to help cheer up the residents in the very gray and white landscape around them.

The winter team at Halley VI includes a chef, a doctor, an electrician, several mechanics, electronics engineers, and a heating and ventilation engineer.

The upper-level climate observatory has a 360-degree view of the ice shelf.

Seven GPS sensors known as the "Lifetime of Halley" monitor the movement of the ice shelf.

Most of the base was made in factories and workshops in the UK and South Africa, then shipped to Antarctica.

The base is named after Edmond Halley, who was a famous 17th–18th century scientist and astronomer.

A colony of emperor penguins keeps the staff company from May until February.

HOW TO BUILD IN THE SEA

We know from what we've learned in this book that one
of the challenges in creating stable foundations
in rivers is that the soil below is wet and soft.

To allow construction in this type of ground, engineers have to move the water
out of the way using special equipment, such as the caissons used for the
Brooklyn Bridge. We also know that water is dense and heavy compared to
air, so when we build structures in water, more material is needed to resist the
pressure and stop the structure from leaking. These challenges grow in water that
is deeper, further from land, subject to strong currents and salty. As technology
has advanced, however, we've been able to build ever-more ambitious structures
underwater, from sea bridges and oil rigs even to restaurants!

Bandra–Worli Sea Link

The Bandra–Worli Sea Link is a beautiful bridge
that joins two peninsulas of land across the
large Mahim Bay in Mumbai, India. It is one
of the longest bridges in India at 3.5 miles
and the first bridge in India to be constructed
in the sea. This bridge is made from tall
concrete columns, or piers, that are about
164 feet apart. To build these columns, deep
foundations had to be dug underwater.

Engineers used jack-up barges, which
are large platforms with long legs that can
stretch right down to the seabed. A piling
rig created the piles. In the meantime,
steel cofferdams were assembled on
land. Cofferdams are watertight enclosures
without a base which are temporary and
are only used during construction. They
were brought out to sea and lowered down
so their lower edges locked into the
ground. Then the water was pumped
out to form a dry place to pour
concrete for the foundation
slab and bridge columns.

ITHAA RESTAURANT

Ithaa, which means "mother-of-pearl" in Divehi, the local language of the Maldives, is one of the most beautiful restaurants in the world. Amazingly, it sits 16 feet below sea level alongside a coral reef on the island of Rangali, in the Maldives. You can watch fish and see vibrant coral all around you while you enjoy a fancy six-course meal. This view is thanks to a big transparent acrylic (a type of plastic) tunnel that resists the pressure of the water.

Since the restaurant is on the inside of the main reef, the forces aren't as strong as they would be in the open sea. But engineers still had to design for the tides rising and falling, the movement of waves, the change in water pressure as the structure was lowered down, and even for the effects of rising sea levels caused by climate change.

The restaurant was constructed in Singapore on land in just five months. The acrylic tunnel is built around a steel skeleton, and the base is filled with concrete to make it heavy so it stays at the bottom of the sea. And it is very carefully sealed to make sure there are no leaks.

To tie the structure down, four huge steel pipe piles were hammered into the seabed. A large barge with its own crane transported the 193-ton structure from Singapore to the Maldives. Divers moved 94 tons of sand in bags into the structure to make it sink down toward the piles. Finally, the structure was attached to the steel piles by pouring in concrete to make sure it wouldn't float away.

HOW TO BUILD IN OUTER SPACE

At the moment, space exploration by humans relies on the International Space Station or spacecraft. But some engineers and scientists are studying how to build structures on the Moon! Then researchers could live there for a long time to carry out their experiments and learn more about outer space.

What's it like on the Moon?
The force of gravity on the Moon is six times less than that on Earth. The amount of force pulling down on objects is smaller, and so a lighter material could be used to build a structure on the Moon than on Earth. It won't float away—even this smaller force of gravity is enough to make sure of that. In past missions to the Moon, lunar landers and rovers didn't need to be anchored to the surface.

The challenges
The lunar environment is even harsher than Antarctica. The temperature on the Moon varies hugely. When the sun is shining on its surface, the temperature reaches 250°F, but then it plummets to around −275°F when it's in darkness. Unlike Earth, which is surrounded by gases that protect it, there is very little atmosphere on the Moon. Not only does this prevent humans from being able to breathe without help, it means the Moon is exposed to cosmic rays and radiation. These are types of energy made of waves and particles that hurtle through space like light. Some types of radiation are very dangerous for humans. The Moon is also very, very dusty, with micro-meteorites, which are tiny space rocks, constantly bombarding its surface. Then there are the "moonquakes" to contend with . . .

DID YOU KNOW?
One possible design for the lunar structure is inspired by the igloo, a structure that originated in parts of the world where nights are very long during the winter and days are very long during the summer. One day on the Moon lasts about 29 Earth days, which means the Sun shines on a point on the Moon for about half that time, and it is in darkness for the other half. Compare this to Earth, where our day is 24 hours long with an average of 12 hours of sun and 12 hours of darkness.

What can we build from?

It makes sense to use materials that are easy to find on the Moon,
such as the top layer of regolith—the loose dust and particles of rock that
cover the surface. There are two ways to turn regolith into a solid material.
You could use a "glue" to bind the grains together, or you could fuse them
by melting the regolith. Scientists are investigating both options. If the glue
method is used, it will form a paste, and this paste could be 3D printed by
robots. This is ideal in a difficult environment because it cuts down on humans
having to do complicated building processes, such as cutting or welding.
But 3D printing works well on Earth because of gravity. Engineers are still
studying what the Moon's low gravity means for printing there.
On Earth, the gravity makes sure that as the layers of material are printed,
they bond together well. With reduced gravity, the layers might not stick
as strongly, so the material could be weaker. The robots would have
the added challenge of needing to withstand the dust, the extreme
temperatures, and the meteorites—things we don't need to
worry about on Earth.

Will they get built?

The answer is: we don't know yet! It depends
on whether the leaders of different countries
decide that long-term exploration of the Moon
is useful. It's an exciting project, and things are
still being designed and tested as we speak.

BUILDING INTO THE FUTURE

Engineering has shaped our whole world: our buildings and bridges, our phones and computers, and even how we travel and live in space!

Imagine what our future cities could look like: pencil-thin soaring towers, underwater houses with glass that can't be shattered, bridges that span more than ten times the distance they do today. All of these ideas could one day become real. Engineers are constantly learning from the world around them and having to adapt to changes in the environment and in society. They are always trying to improve the way we build and challenge ideas of what's possible. Sometimes we are only limited by our imaginations. By using new materials and trying out different methods, engineers will keep finding ways to make ever more incredible structures. If you let your imagination go, what would you build?

MATERIALS

Self-healing concrete

A special concrete has been invented that has tiny capsules mixed in. The capsules contain a type of bacteria—normally found near lakes close to volcanoes—which can survive without food or oxygen for 50 years. If cracks form in the concrete and water seeps in, the capsules release the bacteria. The bacteria then eat the capsules and convert the chemicals to limestone, which fills in the cracks. This type of concrete will be really useful in areas that can't be easily accessed after they've been built, for example deep in the ground.

Aluminum foam

Injecting air into molten aluminum causes pockets to be formed. The metal cools and becomes solid around these air bubbles to create aluminum foam panels. Not only do the panels look great, they are very light and can be 100% recycled. They have been used as cladding to protect buildings from the wind and rain.

Bamboo

Bamboo has been used to build structures for over a thousand years in countries such as China and India. It is very strong and grows very quickly, which makes it environmentally friendly. In Hong Kong, people still use bamboo scaffolding to build skyscrapers. Now engineers are looking to use bamboo as reinforcement in concrete instead of steel.

Carbon nanotubes

A nanometer is a billionth of a meter, which is really, really small. Scientists have created nanotubes made from carbon—they are formed from a sheet of carbon atoms just one nanometer thick and arranged in hexagons, wrapped into a tube shape. These tiny tubes are the strongest material for their weight in the world. Engineers are investigating how these tubes can be embedded into steel, concrete, wood, and glass to make those materials stronger.

METHODS

3D Printing

3D printing can be used to create not only small components for structures, but entire structures. Entirely 3D-printed footbridges have been constructed around the world. From a concrete bridge in China to a steel bridge that was on display at Dutch Design Week, structures made in this way are lighter than conventional structures and quicker to build.

Biomimicry

Engineers are learning from nature. Instead of just copying the shapes of natural structures, we are also being inspired by their function. The skulls of certain birds have two layers of thin bone with a web of connections between the layers woven between large air pockets. This makes them light and strong. By building structures this way, we can make them light and strong too. Engineers have also studied the skeleton of a sea urchin, which is made up of interlocking plates. The Landesgartenschau Exhibition Hall in Stuttgart, Germany, is made from plywood sheets that interlock this way to create an incredibly slender, but very strong, canopy.

Robotics

Inspired by termites that work in groups of hundreds of thousands to build huge mud structures, engineers are designing small robots that are programmed to work together as a swarm. Their sensors detect the work they need to do, the presence of other robots, and rules for getting out of each other's way. They can be programmed to build brick walls, protection against flooding along coasts, or deep underwater pipelines.

Virtual reality

Technology is playing a huge role in the way design and construction are evolving. One such technique is using virtual reality (VR). We know that VR is fun for gaming, but engineers are creating complex worlds in a computer to simulate what their structures will look like. This way, designers and other interested people can wear a headset and walk around the finished virtual structure.

GLOSSARY

3D printing A method of printing—instead of printing with ink to make a flat image, layers and layers of material are built up in the right shape to make a 3D solid object

aggregates Crushed-up pieces of rock, stone, and brick used along with cement and water to make concrete

airtight Completely sealed so that no air can get in or out

algorithm A set of rules or instructions for carrying out tasks, such as particular calculations or steps in math, often done by a computer

appease Do something to stop anger and bring peace and calm

aqueduct A structure built to carry water—for example a canal, tunnel, or bridge

aquifer A layer of rock that holds water underground

atoms Tiny basic particles of matter, from which everything is made. An element contains only one type of atom.

Aztecs North and Central American people who lived in an area spanning much of modern-day Mexico, Belize, and Guatemala from around the 13th to the 15th century

bacteria Microscopic, single-celled living things, some of which cause diseases in humans

bascule A piece of a bridge that can lift up to let ships through

beam A long, strong piece of wood, metal, or concrete used to make the frame that supports a building or bridge

bearing Concentrating something's weight downwards toward the ground, which has the effect of holding the thing up

Bernoulli numbers A particular mathematical number sequence. Ada Lovelace found a way of working it out.

borehole A very deep, round well

caisson A large watertight chamber used for carrying out work underwater

cantilever A type of structure that is supported at one end while the other end is free, like a diving board

carbon An element which exists in several different forms, including diamond, graphite, and coal

column An upright pillar

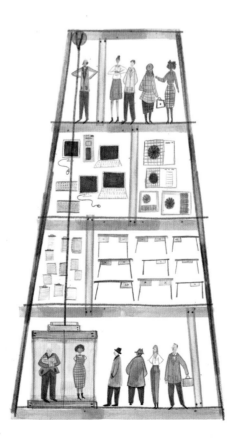

compression A squashing force

concrete A building material made by mixing cement (burned powder of certain rocks), aggregates (crushed-up pieces of rock), and water

crystal A solid in which the atoms or molecules are arranged in a regular pattern. Most metals and many rocks have a crystal structure.

diagrid A supportive frame built on the outside of a structure

embankment A thick wall like an artificial riverbank built beside a river, road, or railway

facade The outside surface or "face" of a building

foundation The part of a structure that is built underground and forms a solid support for the rest of the structure

frequency The number of times something sways or vibrates in one second

friction A force between two surfaces that are sliding against each other—friction grips and tries to stop things from sliding

gravity The force which pulls objects toward each other and pulls us to the Earth

hydration A chemical process in which water reacts with another substance. In the process of making concrete, hydration happens when the burned powder and water are mixed together, and they thicken up and bind together.

hydraulic Powered by water or another fluid under pressure

hydro-power Electricity generated by the movement of water, for example as it rushes over a dam or swells with the tide

indigenous A word used to describe the people who lived in a region first, before people from another part of the world arrived. It is also used to describe plants and animals.

irrigating Watering

jib The arm of a crane

landlocked A country that has no coastline and is completely surrounded by other countries

mesh A structure like a net, made from wire, thread, or plastic, used to strengthen another material

metal A substance that conducts electricity and heat well. Metals are usually strong, shiny, and hard and include iron, steel, aluminum, gold, and zinc.

mineral A substance found naturally in the Earth's surface. Metals are minerals but there are many others, including quartz, mica, salt, lime (calcium oxide), and sulfur.

oculus A circular opening in a building

pendulum A weight that hangs from the end of a wire or string. One end of the wire is fixed and the weight hangs on the other end and can swing freely from side to side.

peninsula A long narrow finger of land that sticks out from a larger area of land and is almost completely surrounded by water

piston A machine made of a cylinder of fluid (such as oil, water, or steam) with a rod inside. The rod moves up and down inside the cylinder and pushes another object, making it move in turn.

pneumatic Filled with compressed air

pontoon A big hollow platform that floats on water, for example to form a bridge

reinforce Strengthen

resonance When something has a force on it which makes it vibrate very strongly at a particular frequency compared with other frequencies, that is resonance. For example, if you push a swing at a particular frequency, or number of times per minute, you'll find that it goes higher than if you try to push it faster or slower.

retractable Something that can be moved inward or backward, such as a roof that rolls back to open

sensor A piece of equipment that detects something such as movement, light, heat, pressure, and so on

sewer An underground tunnel that carries waste water and sewage away

shaft A long vertical tunnel going down into the ground

span Reach across a gap, or a structure that reaches across

spire A tall pointed structure on top of a building

suspension bridge A type of bridge in which the bridge deck is suspended, or hung, from thick cables that stretch from one end of the bridge to the other

tension A force that pulls on objects, stretching them apart

termites Insects, similar to ants, that live together in very large numbers and work together to build giant mound-shaped homes out of soil and dung

terrain The ground. If something is described as difficult terrain, it is difficult to travel across or build on—for example, it might have steep mountains or swamps.

trestle A structure supported by lots of columns

tributary A stream or river that flows into a bigger stream or river

truss A frame made up of lots of smaller beams and columns

virtual reality A type of computer technology that makes someone feel like they are somewhere else. The person wears a headset, and computer images, sounds, and other sensations create the effect.

wrought iron A form of iron that is quite easy to bend and shape but is not as strong as other forms of iron

ENGINEERS' GALLERY

Katie Kelleher is a crane operator. She lifts materials to build tunnels and train stations using different types of cranes. Her team helps her because being high up in a crane means you can't always see what you're lifting off the ground.

Sir Mokshagundam Visvesvaraya was a civil engineer who worked with water. He has universities, museums, and train stations named after him. India celebrates Engineer's Day on September 15 every year in his memory.

Agnes Jones is a blacksmith who makes art out of steel. She heats metals up to 2,200°F to shape them into benches, frames, and even people!

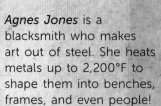

Ilya Espino de Marotta was made the Chief Engineer of the Panama Canal expansion project. She was the first woman ever to take on that role.

Dr. Nike Folayan is an engineer who works on safety in road, rail, and tunnels. She became an engineer because she was fascinated by how TV antennae work!

Bill Baker is a structural engineer who has designed some of the world's tallest towers, including the Burj Khalifa. He and his team developed the "buttressed core" system to keep it stable against wind and earthquakes.

Marcus Vitruvius Pollio was a Roman architect and engineer who wrote a ten-volume book called *De Architectura*. He said that all buildings should have three attributes—strength, utility, and beauty.

Dr. Advenit Makaya is an advanced manufacturing engineer at the European Space Agency. One of the projects he is working on is what materials to build with in space!

Dr. Efrain Ovando-Shelley was one of the geotechnical engineers who helped save the Metropolitan Cathedral in Mexico. He is an expert on how soil behaves and how structures should be built on particular types of ground.

Roma Agrawal is a structural engineer. She worked on the design of The Shard as well as train stations, apartment buildings, and a sculpture. She also writes books.

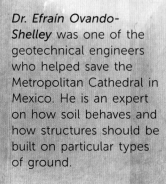

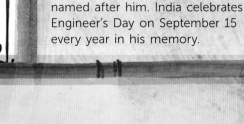

HOW TO BUILD IN OUTER SPACE

At the moment, space exploration by humans relies on the International Space Station or spacecraft. But some engineers and scientists are studying how to build structures on the Moon! Then researchers could live there for a long time to carry out their experiments and learn more about outer space.

What's it like on the Moon?

The force of gravity on the Moon is six times less than that on Earth. The amount of force pulling down on objects is smaller, and so a lighter material could be used to build a structure on the Moon than on Earth. It won't float away—even this smaller force of gravity is enough to make sure of that. In past missions to the Moon, lunar landers and rovers didn't need to be anchored to the surface.

The challenges

The lunar environment is even harsher than Antarctica. The temperature on the Moon varies hugely. When the sun is shining on its surface, the temperature reaches 250°F, but then it plummets to around −275°F when it's in darkness. Unlike Earth, which is surrounded by gases that protect it, there is very little atmosphere on the Moon. Not only does this prevent humans from being able to breathe without help, it means the Moon is exposed to cosmic rays and radiation. These are types of energy made of waves and particles that hurtle through space like light. Some types of radiation are very dangerous for humans. The Moon is also very, very dusty, with micro-meteorites, which are tiny space rocks, constantly bombarding its surface. Then there are the "moonquakes" to contend with . . .

ITHAA RESTAURANT

Ithaa, which means "mother-of-pearl" in Divehi, the local language of the Maldives, is one of the most beautiful restaurants in the world. Amazingly, it sits 16 feet below sea level alongside a coral reef on the island of Rangali, in the Maldives. You can watch fish and see vibrant coral all around you while you enjoy a fancy six-course meal. This view is thanks to a big transparent acrylic (a type of plastic) tunnel that resists the pressure of the water.

Since the restaurant is on the inside of the main reef, the forces aren't as strong as they would be in the open sea. But engineers still had to design for the tides rising and falling, the movement of waves, the change in water pressure as the structure was lowered down, and even for the effects of rising sea levels caused by climate change.

The restaurant was constructed in Singapore on land in just five months. The acrylic tunnel is built around a steel skeleton, and the base is filled with concrete to make it heavy so it stays at the bottom of the sea. And it is very carefully sealed to make sure there are no leaks.

To tie the structure down, four huge steel pipe piles were hammered into the seabed. A large barge with its own crane transported the 193-ton structure from Singapore to the Maldives. Divers moved 94 tons of sand in bags into the structure to make it sink down toward the piles. Finally, the structure was attached to the steel piles by pouring in concrete to make sure it wouldn't float away.